Service Learning and Environmental Chemistry: Relevant Connections

ACS SYMPOSIUM SERIES **1177**

Service Learning and Environmental Chemistry: Relevant Connections

Elizabeth S. Roberts-Kirchhoff, Editor
University of Detroit Mercy
Detroit, Michigan

Matthew J. Mio, Editor
University of Detroit Mercy
Detroit, Michigan

Mark A. Benvenuto, Editor
University of Detroit Mercy
Detroit, Michigan

Sponsored by the
ACS Division of Environmental Chemistry, Inc.

American Chemical Society, Washington, DC

Distributed in print by Oxford University Press

Library of Congress Cataloging-in-Publication Data

Service learning and environmental chemistry : relevant connections / Elizabeth S. Roberts-Kirchhoff, editor, University of Detroit Mercy, Detroit, Michigan, Matthew J. Mio, editor, University of Detroit Mercy, Detroit, Michigan, Mark A. Benvenuto, editor, University of Detroit Mercy, Detroit, Michigan ; sponsored by the ACS Division of Environmental Chemistry, Inc.
   pages cm. -- (ACS symposium series ; 1177)
   Includes bibliographical references and index.
   ISBN 978-0-8412-3008-8 (alk. paper)
   1. Environmental chemistry. 2. Environmental monitoring. 3. Chemistry--Social aspects --Louisiana--New Orleans. 4. Hurricane Katrina, 2005. I. Roberts-Kirchhoff, Elizabeth S., editor. II. Mio, Matthew J. (Matthew John), 1974- editor. III. Benvenuto, Mark A. (Mark Anthony), editor. IV. Xavier University of Louisiana.
   TD193.S475 2014
   577'.140715--dc23
                                           2014046393

The paper used in this publication meets the minimum requirements of American National Standard for Information Sciences—Permanence of Paper for Printed Library Materials, ANSI Z39.48n1984.

Copyright © 2014 American Chemical Society

Distributed in print by Oxford University Press

All Rights Reserved. Reprographic copying beyond that permitted by Sections 107 or 108 of the U.S. Copyright Act is allowed for internal use only, provided that a per-chapter fee of $40.25 plus $0.75 per page is paid to the Copyright Clearance Center, Inc., 222 Rosewood Drive, Danvers, MA 01923, USA. Republication or reproduction for sale of pages in this book is permitted only under license from ACS. Direct these and other permission requests to ACS Copyright Office, Publications Division, 1155 16th Street, N.W., Washington, DC 20036.

The citation of trade names and/or names of manufacturers in this publication is not to be construed as an endorsement or as approval by ACS of the commercial products or services referenced herein; nor should the mere reference herein to any drawing, specification, chemical process, or other data be regarded as a license or as a conveyance of any right or permission to the holder, reader, or any other person or corporation, to manufacture, reproduce, use, or sell any patented invention or copyrighted work that may in any way be related thereto. Registered names, trademarks, etc., used in this publication, even without specific indication thereof, are not to be considered unprotected by law.

PRINTED IN THE UNITED STATES OF AMERICA

Foreword

The ACS Symposium Series was first published in 1974 to provide a mechanism for publishing symposia quickly in book form. The purpose of the series is to publish timely, comprehensive books developed from the ACS sponsored symposia based on current scientific research. Occasionally, books are developed from symposia sponsored by other organizations when the topic is of keen interest to the chemistry audience.

Before agreeing to publish a book, the proposed table of contents is reviewed for appropriate and comprehensive coverage and for interest to the audience. Some papers may be excluded to better focus the book; others may be added to provide comprehensiveness. When appropriate, overview or introductory chapters are added. Drafts of chapters are peer-reviewed prior to final acceptance or rejection, and manuscripts are prepared in camera-ready format.

As a rule, only original research papers and original review papers are included in the volumes. Verbatim reproductions of previous published papers are not accepted.

ACS Books Department

Contents

1. Service Learning, Chemistry, and the Environmental Connections 1
   Elizabeth S. Roberts-Kirchhoff, Mark A. Benvenuto, and Matthew J. Mio

2. Environmental Justice in New Orleans and Beyond: A Freshman Seminar Service-Learning Course at Xavier University of Louisiana 5
   Michael R. Adams

3. Green Action Through Education: A Model for Fostering Positive Attitudes about STEM by Bringing the Scientists to the Students, Not the Other Way Around 23
   William J. Donovan, Ethel R. Wheland, Gregory A. Smith, and Angela Bilia

4. Connections between Service Learning, Public Outreach, Environmental Awareness, and the Boy Scout Chemistry Merit Badge 67
   Mark A. Benvenuto and Matthew J. Mio

5. A Service-Learning Project Focused on the Theme of National Chemistry Week: "Energy-Now and Forever" for Students in a General, Organic, and Biological Chemistry Course 73
   Elizabeth S. Roberts-Kirchhoff

6. Service Learning in Environmental Chemistry: The Development of a Model Using Atmospheric Gas Concentrations and Energy Balance To Predict Global Temperatures with Detroit High School Students 87
   Daniel B. Lawson

7. Environmental Justice through Atmospheric Chemistry 105
   Nicole C. Bouvier-Brown

8. Using Service Learning To Teach Students the Importance of Societal Implications of Nanotechnology 123
   A-M. L. Nickel and J. K. Farrell

9. XRF Soil Screening at the Alameda Beltway 135
   Steven Jon Bachofer

10. Bottled Water Analysis: A Tool For Service-Learning and Project-Based Learning 149
    Olujide T. Akinbo

11. Instrumental Analysis at Seattle University: Incorporating Environmental Chemistry and Service Learning into an Upper-Division Laboratory Course 193
    Douglas E. Latch

12. The LEEDAR Program: Learning Enhanced through Experimental Design and Analysis with Rutgers 209
    David A. Laviska, Kathleen D. Field, Sarah M. Sparks, and Alan S. Goldman

Editors' Biographies 229

Indexes

Author Index 233

Subject Index 235

Chapter 1

Service Learning, Chemistry, and the Environmental Connections

Elizabeth S. Roberts-Kirchhoff,* Mark A. Benvenuto, and Matthew J. Mio

University of Detroit Mercy, Department of Chemistry and Biochemistry, 4001 W. McNichols Road, Detroit, Michigan 48221-3038
*E-mail: robkires@udmercy.edu.

An overview of service learning is presented, with emphasis on how service learning applies in college-level chemistry courses, and how environmental chemistry and awareness can be related to the area of community service.

Introduction

The professor stands, as professors and lecturers have done in countless places at countless times since the founding of the University of Bologna in the year 1088, and talks. And talks. And talks some more. At the best, students are enraptured by the lecture, and leave smarter than when they walked into the lecture hall. At the worst, they fall asleep or are unmotivated to return to the next course meeting.

Lecturing in the college classroom may be one of the biggest continuing paradoxes of higher education: it can work very well and be effective, and apparently has done so for nearly a millennium. But the exact same model and delivery method can be completely ineffective. One can imagine that professors and teachers have been trying to improve the learning experience almost since its inception. Numerous tools have been developed in attempts to make learning a richer, more meaningful experience. Freeman and colleagues performed a meta-analysis of 225 different studies on the impact of active learning techniques or the traditional lecture format on students' test scores and failure rates in undergraduate science, technology, engineering and mathematics (STEM) courses (*1*). In courses with some format of active learning, the average test score increased by 6 %. In addition, the students in traditional lecture courses were more likely to fail than in a course with active-learning techniques (*1*). Active learning strategies can include collaborative problem solving, personal response systems, studio courses, ConcepTests, case method teaching, and service learning. Ten active-learning strategies were reviewed by Kuh as high-impact educational practices (*2*). These were defined as high impact since participation in these activities resulted in increases in student engagement, grade point averages, and retention. Most importantly, all students benefited in some way and there was a higher impact for historically underserved students. The high-impact active-learning strategies included first-year seminars, common intellectual experiences, learning communities, writing-intensive courses, collaborative projects, undergraduate research, global learning, service learning, internships, and capstone projects (*2*).

One of these active-learning activities, service learning, attempts to increase learning, and at the same time engage students with their greater community (*3, 4*). Service learning is generally defined as a combination of active learning and practice, most often serving the community. In the strongest cases, service learning is coupled with substantive reflection, making obvious connections to the role a university plays in society, and sometimes fulfilling a religious mission. Service learning has in many cases deepened such connections. In addition, depending on how the service learning is structured, it may involve more than one of the ten high-impact pedagogical practices (*1*). That is, the service-learning activity could also involve a learning community, a collaborative project, or global learning especially with regard to civic responsibility or social justice issues.

Service learning is more, however, than simply active participation in some local activity or event and a connection to class work. There is almost always some form of reflection on the part of the students to help connect what they have done with the course content or on their personal growth. This is one of the reasons that service learning has been utilized as long and as widely as it has in liberal arts courses (*3, 4*). Numerous composition classes, as well as classes on such topics as ethics and religious studies, incorporate some form of reflection, and service learning provides the experience upon which to draw. Thus, to many, it seems that service learning is confined to the liberal arts, simply because it is straightforward to add reflection to the traditional disciplines of the humanities.

Environmental Connections and Incorporation of Service Learning into Chemistry Courses

The science and engineering disciplines have obvious connections to community involvement that lend themselves to the use of service learning. While there are several broad thrusts for chemistry professors who wish to use such techniques, two that have become rather common are: utilizing college students as teachers or mentors to younger students in the greater community (*5–11*); and second, performing some form of local environmental monitoring or analysis in a chemistry course (*12–19*). The science and engineering link to environmental awareness may go back as far as the publication of Rachel Carson's *Silent Spring*

(*20*). Chemists have found that soil, water, and air monitoring and analyses can serve as excellent connections to their course work, and that service learning is a very good tool to make these connections (*19*).

Like many techniques and tools, service learning within chemistry courses is a way to help students supplement their education with a lasting appreciation for the subject matter they have learned. Perhaps the reason the traditional lecture continues to be a mainstay of higher education is because the students, faculty, and administrators are all very used to it. Adding a service-learning component to many chemistry courses may not be difficult, though. While it requires extra time outside the lecture hall or teaching laboratory, this increase appears to be more than offset by the depth of student learning, and by what experiences they take away from the course. The learned connections can be as varied as the faculty members who teach different chemistry lecture or lab courses, or who engage in associated research projects (*5–19*).

Summary

There are many ways in which service learning can be incorporated into the chemistry curriculum, whether in the traditional lecture class, in laboratory classes, or in research projects. Environmental chemistry connections are also unmistakable, and serve as studies that are both practical and of concern to our students and other members of our communities. All such techniques and scenarios can enrich a student's learning experience, and be beneficial to some segment of the local population as well.

In short, service learning can help move the study of chemistry away from a model that has in some way been practiced since the year 1088 — one might say away from all the Bologna — to meaningful interactions with our students that involve both learning and raising awareness of conditions in their shared environment. Service learning, therefore, is a powerful tool by which chemists can effect both pedagogical and social change.

References

1. Freeman, S; Eddy, S. L.; McDonough, M.; Smith, M. K.; Okoroafor, N.; Jordt, H.; Wenderoth, M. P. Active Learning increases Student Performance in Science, Engineering, and Mathematics. *Proc. Natl. Acad. Soc. U.S.A.* 2014 DOI: 10.1073/pnas.1319030111.
2. Kuh, G. D. *High-Impact Educational Practices: What are they, who has access to them, and why they matter*; Association of American Colleges and Universities: Washington, DC, 2008.
3. Bringle, R. G.; Phillips, M. A.; Hudson, M. *The Measure of Service Learning: Research Scales to assess Student Experiences*; American Psychological Association: Washington DC, 2003.
4. *National Service Learning Clearinghouse*; http://gsn.nylc.org (accessed 23 August 2014).

5. Kalivas, J. H. A Service-Learning project based on a Research Supportive Format in the General Chemistry Laboratory. *J. Chem. Educ.* **2008**, *85*, 1410–1415.
6. Sutheimer, S. Strategies to Simplify Service-Learning Efforts in Chemistry. *J. Chem. Educ.* **2008**, *85*, 231–233.
7. Harrison, M. A.; Dunbar, D.; Lopatto, D. Using Pamphlets to Teach Biochemistry: A Service-Learning Project. *J. Chem. Educ.* **2013**, *90*, 210–214.
8. Esson, J. M.; Stevens, R.; Thomas, A. Service-learning in Introductory Chemistry: Supplementing Chemistry Curriculum in Elementary Schools. *J. Chem. Educ.* **2005**, *82*, 1168–1173.
9. Glover, S. R.; Sewry, J. D.; Bromley, C. L.; Davies-Coleman, M. T.; Hlengwa, A. The Implementation of a Service-Learning Component in an Organic Chemistry Laboratory Course. *J. Chem. Educ.* **2013**, *90*, 578–583.
10. LaRiviere, F. J.; Miller, L. M.; Millard, J. T. Showing the true face of Chemistry in a Service-Learning Outreach Course. *J. Chem. Educ.* **2007**, *84*, 1636.
11. Morgan-Theall, R. A.; Bond, M. R. Incorporating Professional Service as a component of Genenal Chemistry Laboratory by demonstrating Chemistry to Elementary Students. *J. Chem. Educ.* **2013**, *90*, 332–337.
12. Gardella, J. A.; Milillo, T. M.; Gaurav, O.; Manns, D. C.; Coeffey, E. Linking Advanced Public Service Learning and Community Participation with Environmental Analytical Chemistry: Lessons from case studies in Western New York. *Anal Chem.* **2007**, *79*, 811–818.
13. Fitch, A.; Wang, Y.; Mellican, S.; Macha, S. Lead Lab Teaching Instrumentation with One Analyte. *Anal Chem.* **1996**, *68*, 727A–731A.
14. Kammler, D. C.; Truong, T. M.; VanNess, G.; McGowin, A. E. A Service-Learning Project in Chemistry: Environmental Monitoring of a Nature Preserve. *J. Chem. Educ.* **2012**, *89*, 1384–1389.
15. Burand, M. W.; Ogba, M. Service-Learning General Chemistry: Lead Paint Analyses. *J. Chem. Educ.* **1999**, *76*, 920.
16. Draper, A. J. Integrating Project-Based Service-Learning into an Advanced Environmental Chemistry Course. *J. Chem. Educ.* **2004**, *81*, 221.
17. Juhl, L.; Yearsley, K.; Silva, A. *J. Chem. Educ.* **1997**, *74*, 1431.
18. Kesner, L.; Eyring, E. M. Letter Writing as a Service-Learning Project: An Alternative to the Traditional Laboratory Report. *J. Chem. Educ.* **2013**, *90*, 1701–1702.
19. Ward, Harold, Ed. Acting Locally: Concepts and Models for Service-Learning in Environmental Studies. AAHE's Series on Service-Learning in the Disciplines. *American Association for Higher Education*, Sterling, VA 1999.
20. Carson, R. *Silent Spring*; Houghton Mifflin: Boston, 1962.

Chapter 2

Environmental Justice in New Orleans and Beyond: A Freshman Seminar Service-Learning Course at Xavier University of Louisiana

Michael R. Adams**

Department of Chemistry, Xavier University of Louisiana, 1 Drexel Drive, New Orleans, Louisiana 70125
*E-mail: mradams@xula.edu.

The Freshman Seminar program at Xavier University of Louisiana has as its foundation the university's mission to contribute to the promotion of a more just and humane society. Relying on their own academic discipline, instructors identify a theme and structure their course to include service learning related to the mission. Partnering with a local environmental action organization, students work in communities that border refineries and chemical plants in the area along the Mississippi River sometimes referred to as Cancer Alley. Projects involve surveying residents regarding health issues, raising awareness of how to stay informed when accidents occur, and informing residents of opportunities for reporting possible releases from neighboring facilities. Classroom activities are designed to increase students' ability to critically interpret reports of chemical accidents. Through this experience, students gain a deeper understanding of issues of environmental justice.

Introduction

In August 2005 the Xavier University of Louisiana campus and community were devastated by the effects of Hurricane Katrina. However, the students, faculty, and staff were determined to rebound from this tragedy and return as a stronger academic community. In moving forward, we looked closely at many of our programs, including the overall First Year Experience (FYE) for new students. A number of faculty and administrators had a strong desire to develop as part of the FYE program a new freshman seminar sequence, one that had a more academic focus than the previous non-credit courses that were required of all freshmen.

Faculty representing a wide variety of disciplines gathered to devise a plan for developing these new courses. What was clear early in the process was that the course sequence would be mission-based and that service learning would be a required component. The planning group agreed that service learning would be most appropriate for helping students to understand the mission of the University, stated below:

"Xavier University of Louisiana, founded by Saint Katharine Drexel and the Sisters of the Blessed Sacrament, is Catholic and historically Black. The ultimate purpose of the University is to contribute to the promotion of a more just and humane society by preparing its students to assume roles of leadership and service in a global society. This preparation takes place in a diverse learning and teaching environment that incorporates all relevant educational means, including research and community service."

What remained a challenge was how to incorporate different disciplines and faculty from different departments, while still maintaining common goals for all sections of the course. Ultimately, this led to a decision to ask faculty who were part of the instructional team to bring their own expertise and discipline to the teaching of the course by developing a unique theme in partnership with a community group with whom they would develop a service-learning program. The guiding principle is that the theme should be consistent with the mission of the University, with connections to the overall concept of social justice.

As a chemist with no service learning experience and whose entire 20+ year teaching experience had been in chemistry, the task of developing a course centered on a service-learning project and theme that involved chemistry but was not specifically designed for science majors was daunting. However, a chance meeting with representatives from the Louisiana Bucket Brigade, a non-profit grassroots environmental action organization, has led to development of a successful and well-received course that is frequently cited as a model for new faculty to follow in developing their own freshman seminar courses. Student opinion has been overwhelmingly positive and the relationship with our community partner has remained strong.

Course Sequence Description

As a result of the post-Katrina discussions and planning described above, the Freshman Seminar program at Xavier has evolved to become a two-semester sequence. All new students (freshmen and transfers with fewer than 30 credit hours) are required to complete the sequence as part of the core curriculum. Each semester consists of a one-credit course with twelve 50-minute class meetings per semester. Course instructors are required to teach both semesters of the sequence, working with the same group of students each semester. In a typical year, faculty from 10-15 different academic departments volunteer to be part of the instructional team.

The first-semester course, FRSM 1000, includes coverage of many topics typically seen in first-year-experience courses, with a heavy focus on college survival skills. However, an important component is a common reading that is introduced to students in the summer prior to their arrival on campus. Each year the assigned reading is selected by the campus community, with the requirement that it be appropriate for generating discussions of concepts of social justice. Individual instructors use a common course syllabus as a guide for developing their own unique course, but the common syllabus requires that a specified number of written and oral communication assignments be included. Students in all sections are required to write informal responses to specific questions related to the common reading and many of these writings are in the format of public blogs hosted by websites such as Wordpress.

The second-semester course, FRSM 1100, allows for development of a more unique and focused theme. Although a common syllabus is again used as a guide, instructors are free to choose their own readings and develop their own assignments. The course goals are tailored to fit more general goals described in the common syllabus. For the specific course described here, these goals are that students completing the course will have:

- Critically examined the theme of "Environmental Justice: New Orleans and Beyond", especially in relation to specific New Orleans neighborhoods and surrounding communities.
- Further developed their writing and speaking skills, in particular as directed toward various audiences.
- Directly explored an involvement of self in relation to community.
- Demonstrated Xavier's mission to promote a more just and humane society through a service-learning project with the Louisiana Bucket Brigade.
- Utilized competence in technological applications to communicate with various audiences.

It is in FRSM 1100 that service learning is a required component, and faculty are given full support of a dedicated service-learning faculty liaison in our Center for the Advancement of Teaching (CAT), as well as staff from our Center for Student Leadership and Service (CSLS) who are instrumental in helping with the logistics of planning a service project. Support from staff in these two offices has been critical to the overall success of the freshman seminar program, and, specifically, to the course described here. In addition to organizing numerous workshops and seminars, the CAT Faculty in Residence for Service Learning provides direct assistance to instructors in developing course themes and service-learning projects. This is especially helpful to faculty who are new to service learning and are seeking guidance in developing classroom activities and assignments. The Director and staff of the CSLS often help faculty identify appropriate community partners, many of which already have existing relationships with other faculty on campus. When necessary, they also organize transportation and meals for students.

Sections are not discipline-based, but faculty are asked to relate their themes to their own academic disciplines. Because each instructor will be with the same group of students from the first semester course, development and exploration of the theme can grow from the foundation that has been built earlier.

Students are randomly assigned to sections, with efforts made to balance enrollment in terms of major, geographic origin, gender, and academic background. Xavier has a long history of a strong premedical program, and with one of only two pharmacy schools in Louisiana, the prepharmacy program (housed in the Chemistry Department) has a high enrollment. Thus, some 78% of incoming students are intending to major in STEM disciplines, with Business and Psychology being two other popular choices. About half of our students hail from Louisiana, with representation from a total of 40 states and about a dozen foreign countries. The university is an historically Black institution, but our doors are open to all. Approximately 78% of incoming first-year students are African American, with another 8% being Asian and 4% White. Typically, 70+% of students are female. Although admission is selective, approximately 20-25% of first-year students will enroll in at least one developmental course, most commonly math. Typically, each section of Freshman Seminar has an enrollment that mirrors these statistics.

Environmental Justice: New Orleans and Beyond

Development of Course Theme

As described above, the search for a semester theme and service-learning project that drew on my background as a chemist was challenging. However, I was fortunate several years ago to attend a seminar at which I learned of the Louisiana Bucket Brigade and I was introduced to the concept of environmental justice. I developed an immediate interest and desire to further explore this concept, and it seemed that working this into my freshman seminar course was a perfect way to do so. As a result, the following statement is now included in the syllabus and sets the stage for our semester's work:

"Semester Theme: <u>Environmental Justice: New Orleans and Beyond</u> This semester we will examine the relationship between race, class, and environmental issues. We will connect Xavier's mission to environmental justice issues in the New Orleans area and beyond. We will partner with the Louisiana Bucket Brigade, and the central activity for the semester will be a service-learning project completed in conjunction with our community partner."

In order to begin the discussion of the concept of environmental justice, students are asked to submit their own description of what they believe environmental justice to be. Typical answers to this initial question include:

- "In my opinion, the definition of environmental justice is the promotion of a better environment with the enforcement of environmental laws and the participation of every person no matter their race or origin."
- "Environmental Justice can be defined as the protection and cultivation of the natural resources of earth."
- "Environmental Justice to me is when everyone co-operates together to protect the environment and to follow environmental laws. It also has to do with the quality of our environment. No health hazards. It is pretty much when anyone can enjoy their environment no matter where they are."

An open classroom discussion ensues and invariably leads to mentions of global warming, ozone, water quality, recycling, and other typical environmental issues. The understanding of and interest in environmentalism is typical for first-year college students, but it consistently evident that they lack an understanding or knowledge of *environmental justice*. We spend much of the remainder of the semester exploring this concept and how it relates to but is different from environmentalism.

It is at this point that students are introduced to more formal definitions of environmental justice and they begin to see the real focus of the course. According to the U.S. Environmental Protection Agency (*1*), environmental justice is "The fair treatment and meaningful involvement of all people regardless of race, color, sex, national origin, or income with respect to the development, implementation and enforcement of environmental laws, regulations, and policies." Similar other definitions can be found from other sources (*2*), e.g., "The pursuit of equal justice and equal protection under the law for all environmental statutes and regulations without discrimination based on race, ethnicity, and/or socioeconomic status."

In order to facilitate discussion of this new concept, students are asked to read the thought-provoking paper (*3*) by Dr. Deborah Robinson, "Environmental Racism: Old Wine in a New Bottle". Through the discussion that follows, it is clear that the vast majority of students in the class have never considered the connection(s) that may exist between environmental issues and issues of race. The paper provides a number of examples of what can be considered by some to be environmental racism, including the situation in Mossville, LA. Mossville is a predominantly African-American community, founded by free slaves, located in Southwest Louisiana in close proximity to a number of industrial facilities in the greater Lake Charles area. Although Robinson cites a number of issues regarding toxic pollution problems in Mossville that were widely reported in the 1990s, problems still continue today. Any search for news articles mentioning Mossville will invariably produce stories (*4*) dominated by issues of environmental quality, health issues, and the uneasy relationship between Mossville residents and local industry. Recent news (*5, 6*) has been dominated by plans for Sasol to greatly expand its presence in the Lake Charles/Mossville area.

This early discussion of environmental justice sets the stage for our continued study of specific cases throughout the semester, but care is taken to ensure that students are not led to believe that all agree with the opinions that are expressed in some of our early readings. In a Project 21 New Visions Commentary Editorial

(7), David Almasi writes " 'Environmental justice' is a term green activists use to demonize businesses and complain that the government isn't doing enough to help minorities. Their premise is simple: They believe businesses are using political power to unfairly put polluting factories predominately in minority neighborhoods." One of the goals is to help students to be critical readers and to differentiate fact from opinion, especially when reading from sources found through internet searches. Additionally, students are challenged to understand and respect opinions that may differ from their own. A common classroom technique that has been used to accomplish this is a debate in which students are forced to defend positions to which they are assigned. A more detailed description of one of these activities is included later.

Community Partner: Louisiana Bucket Brigade

Concurrent with this initial exploration of environmental justice, students are introduced to our community partner organization for the semester. The Louisiana Bucket Brigade (LABB) is a local non-profit organization dedicated to working with residents of Louisiana who live in close proximity to the numerous oil refineries, chemical plants, and other such industrial facilities found in our state. As stated on the LABB website (http://www.labucketbrigade.org), "The Louisiana Bucket Brigade uses grassroots action to create an informed, healthy society with a culture that holds the petrochemical industry and government accountable for the true costs of pollution *(8)*." This strong partnership with the LABB has been critical to the success of the course over the past several years.

Southeastern Louisiana, especially in the area along the Mississippi River between New Orleans and Baton Rouge, has one of the highest concentrations of petrochemical industrial facilities to be found in the U.S. A total of 17 oil refineries are located in Louisiana and 150 chemical plants are located between New Orleans and Baton Rouge. Some 60,000 oil and gas wells and 36,000 miles of pipeline can be found in the state. On average, there are 10 petrochemical accidents reported per week in Louisiana *(8)*. Due to the high concentration of industrial facilities and the perceived contribution of these facilities to the poor health of residents in parishes along the river, this region is sometimes referred to as "Cancer Alley" *(9)*.

These industrial facilities are interspersed with several small communities, many of which are predominantly African American. Several of the facilities are situated on lands formerly occupied by plantations, and many of the African-American communities grew from settlements of freed slaves and their descendants in the post-Civil War era. Stories of such communities (e.g., Morrisonville, Diamond) being bought out or relocated by large corporations are numerous, as are personal stories of the sense of a loss of history when such communities disappear from the map.

The Louisiana Bucket Brigade gets its moniker from a central service function of the organization, use of the "bucket". The bucket is an easy-to-use device that allows for rapid collection of air samples. Citizens are trained to use the sampling bucket to quickly collect air samples when foul odors emanating from nearby chemical facilities or refineries are noticed. Samples are then sent for analysis and the LABB gathers and monitors this data. The ability of residents to collect samples is especially key in situations where government agencies are likely to respond slowly, if at all, when concern is expressed regarding possible chemical releases.

The bucket is only one of numerous services provided by the LABB. They work closely with residents of many communities to monitor environmental quality and to track releases from industrial facilities. A key effort, one in which students in the course are directly involved, is to encourage residents to report possible releases so that this information can be made public. Reports are summarized on the iWitness Pollution Map (*10*), which allows for a quick view of locations of citizen reports, including the ability to rapidly see hotspots of activity. The Bucket Brigade staff and founding director Anne Rolfes are tenacious in their efforts to hold industry accountable, and more often than not they are contacted by news agencies to provide input for reports related to industrial accidents in Louisiana. They are strong advocates for some who feel they have no voice.

An important and helpful service provided by the LABB is their database (*11*) of chemical releases and other industrial accidents in Louisiana. As an example of some of the information to be found in the database posted on their website, consider the following small sets of data shown in Tables 1 and 2 (LEDQ = Louisiana Department of Environmental Quality):

Table 1. Refinery Accidents Reported to LDEQ 2005-2013

Refinery	City	Total	Pounds	Gallons
Chalmette Refining	Chalmette	539	7,062,996	12,947,169
Marathon Ashland	Garyville	217	792,429	12,610,055
Motiva Enterprises	Norco	246	3,347,620	19,639
Shell Facility	St. Rose	21	913,798	26,662

"Pounds" refers to air emissions while "Gallons" refers to water and ground emissions. "Total" refers to the total number of accidents reported.

Table 2. Chemical Plant Accidents Reported to LDEQ 2005-2013

Chemical Plant	City	Total	Pounds	Gallons
Albemarle PD Corp	Baton Rouge	3	629	0
ExxonMobil Chem.	Baton Rouge	311	5,989,966	130,001,340
Honeywell	Baton Rouge	30	131,545	0
Shell Chem. East	Norco	41	895,925	0

"Pounds" refers to air emissions while "Gallons" refers to water and ground emissions. "Total" refers to the total number of accidents reported.

On the same website the LABB maintains a similar database showing total emissions, categorized by type of pollutant, for 28 industrial facilities located primarily between New Orleans and Baton Rouge.

An early assignment for students is to review the LABB website and write a short piece about anything they discover on the website that they find interesting. Students post their responses on a public blog (*12*), and while their writings cover a broad variety of topics, several will focus on information regarding these industrial accidents.

Course Readings and Other Assignments

Many of the reading and writing assignments early in the semester center on environmental justice issues in Louisiana. Such issues are numerous and assignments vary significantly from year to year. Much of our classroom time is spent openly discussing and debating the issues raised in these reports, with an expansion to more national and global cases later in the term.

One such issue is that of Shintech, Inc., the largest producer of polyvinyl chloride in the U.S. In the mid-1990s Shintech announced plans to build a facility in Convent, LA, a predominantly African-American and low-income community along the Mississippi River about 30 miles west of New Orleans. Citizens spoke out against the plan and a court challenge based on the Environmental Justice Act of 1994 was filed. Eventually Shintech abandoned plans for the facility and chose to construct a smaller facility in Addis, LA, a somewhat more affluent community (*13*).

Not unexpectedly, most students in the class express support for the citizens of Convent and their success in winning their battle. However, students are then asked to look at PVC a bit more closely and are asked the simple question "Are you willing to live without PVC?". We spend some time looking at safety and toxicity information for chemicals associated with the production of PVC and some of the waste products that result. Opinions change quickly and this leads to a discussion of how we, as a society, can ensure that the products we desire as consumers can be produced safely and remain affordable, while protecting the health of people.

A second, more recent situation involves the Motiva Refinery facilities in Norco, LA and Convent, LA. In 2012 Motiva was fined $500,000 for a variety of safety violations. It was concurrently reported (*14*) that Motiva exceeded allowed emission levels for a variety of chemical substances, including sulfur dioxide, sulfuric acid, and volatile organic compounds. Classroom discussion focuses on the effect of such fines and appropriate involvement of government in regulating such industries.

One of the subtle goals of the course is to help students to become more informed citizens and critical thinkers. It is not uncommon for those of us who are trained as scientists and, specifically, chemists to sometimes roll our eyes at what we read in the news reports about chemical "accidents" and the response of citizens, but we also know that we are properly trained to have appropriate respect for the materials we handle. The Motiva situation provides an excellent example; all of the chemicals listed in the report seem equally hazardous to average citizens.

A homework assignment aimed at helping students learn how they can become more informed involves the chemicals listed in Table 3 below, all of which have appeared in recent news articles about industrial accidents in Louisiana:

Table 3. Industrial Chemicals for MSDS Assignment

Titanium Tetrachloride	Hydrochloric acid	Hydrogen Sulfide
Sodium Chromate	Sulfur Dioxide	Benzene
Fluoranthene	Naphthalene	Phenanthrene
Pyrene	1,3-Butadiene	m-Xylene
2-Ethyl-1-hexanol	1,3-Cyclopentadiene	Ethylbenzene
4-bromofluorobenzene	Vinyl acetate	Ammonia
Hydrofluoric acid	Sodium hydroxide	Toluene
Sulfuric acid	Methyl ethyl ketone	Vinyl acetate

Each student is assigned one of these substances and they are asked to find on-line a (Material) Safety Data Sheet for their particular substance. Students are instructed to be prepared to provide an oral summary of their findings in class. In class, students are randomly selected to give brief reports on their assigned chemicals. They are then asked to consider what facts of a chemical release incident they think should be reported in the news and how such information would help them reach their own conclusion about how dangerous the situation might be. They are very good at identifying two important factors: what quantity of the chemical has been released and in what form was it released? Despite their relative inexperience with chemistry, they are also fairly adept at recognizing chemicals that many of us would agree deserve special attention, with many agreeing that hydrofluoric acid seems particularly nasty.

A significant component of the graded work assigned to students are weekly informal writings of at least 300 words that are posted by each student on a public blog website (*12*). Not all of these are necessarily related to the semester theme, but those that are will frequently be in response to short reading assignments. The content of students' blog posts often provides a basis for our open classroom or small group discussions.

These assignments are varied each year, but specific writing prompts from Spring 2014 included the following:

- Prompt: "What are your thoughts on the Obama administration's effort to address environmental justice issues?"
  This assignment was related to a short press release (*15*) that summarized a forum organized in 2010 with a focus on federal government involvement in environmental justice issues. The forum was attended by the secretaries of the Interior, Energy, Health and Human Services,

Homeland Security, and Labor, as well as EPA Administrator Lisa Jackson, who grew up in the New Orleans area. As part of the assignment, students also read a more recent blog post (*16*) regarding White House efforts to address issues of environmental justice. Student responses provided a basis for a class discussion of what the appropriate level of government involvement in these issues ought to be.

- Prompt: "Write a short summary of a news article related to a specific environmental justice issue. Be sure to cite the source of the news article." This particular assignment is recycled each year. Oftentimes, one student's response will be selected for discussion in class and frequently we will build a structured classroom debate around the incident in the news article.

- Prompt: "Basing your comments on the Wallace article (*17*), do you agree that the BP oil spill is an example of an environmental injustice?"

  The 2010 BP Deepwater Horizon oil spill in the Gulf of Mexico is, perhaps, the one industrial environmental disaster most familiar to students in the class. The author draws his title from the widely-reported statement by BP Chairman Carl-Henric Svanburg "we care about the small people" that followed the accident. Although this doesn't seem to fit what one might normally think of as an environmental justice issue, Wallace makes the case that it is and students are asked to provide their own opinions about this. This reading and students' postings provide the foundation for a classroom discussion, one of the more balanced of the semester in terms of variety of opinions expressed. Also, Wallace's commentary includes a specific focus on local citizens of Southeast Asian descent who are engaged in fishing in the gulf. Vietnamese students from the greater New Orleans area make up a significant portion of the non-African American student population at Xavier. Additional mention is made of Native Americans in Louisiana who have been affected by the spill. Students begin to broaden their understanding of environmental justice situations beyond those that affect primarily African-American neighborhoods.

- Prompt: "In what ways do you think this executive order can or will be effective? In what ways does it fall short? If you were to advise President Obama on this issue, what would you tell him?"

  This prompt refers to Executive Order 13650 (*18*), issued by President Obama on August 1, 2013. The order specifically addresses chemical facility safety and security, and, among other things, establishes a working group charged with devising a plan for improved coordination of federal, state, local, and tribal authorities in overseeing chemical facility safety and first response efforts.

Students are also required to write two formal 2-page papers each semester. The second of these assignments will be described later, but the first is to be written in the form of a letter to a government official regarding a specific environmental justice issue. Students are encouraged to use the news article about which they've written in an earlier blog post as the basis for this assignment. Some of the specifics of the assignment are that they must be addressing an appropriate government official, they must support their arguments with appropriate data and sources, and they must be specific in their suggestions for action. While some students will always choose the President as their recipient and others will write rambling, unfocused letters, a good number do an excellent job of writing pointed and detailed letters to lower-level officials who would be more likely to actually receive and read the letter, if sent.

Additional Classroom Activities

Following the activities described above, we spend much of our time in class delving into specific environmental justice issues, all with varying aspects. We expand our discussion of location of refineries and chemical plants to include waste disposal facilities, of which there are several with environmental justice aspects.

One of the more popular annual classroom activities is a structured debate. Using a specific situation of plans for construction of an industrial facility in a particular area (e.g., the Shintech plans for Convent, LA) as a basis, students are assigned such roles as government official, industry representative, parents of local schoolchildren, unemployed citizens, etc.. Working in teams, they develop position statements and must defend their assigned position. Teams are allowed to rebut the statements of other groups, and most do a fairly good job of remaining in their roles. This activity is usually scheduled around midsemester and really allows for students to be brought back to the understanding that these situations are multifaceted and that not all will view them in the same light.

The earlier MSDS activity leads to continued discussion of the vague question "How bad is bad?". Several discussions center on what one's personal response to a chemical accident ought to be and what that of local officials and media should be. Specific attention is given to how to ensure that one has appropriate knowledge to react accordingly when an accident is reported.

A common theme throughout the semester is reliability of sources of information. Readings vary from news reports and editorials to press releases and blogs, so each class period we will invariably have a short discussion about reliability of sources and how to judge this. Separating fact from opinion is a tough challenge for students, but they make good progress each year.

Community Project

Recent community work completed by students centers on the Louisiana Bucket Brigade's iWitness Pollution Map (*10*). The map tracks reports by citizens of flares, foul odors, and other possible chemical releases from industrial facilities and provides a quick method of determining if incidents are isolated or widespread.

Representatives from the Bucket Brigade use two 50-minute class sessions to train class members for their fieldwork. Students visit neighborhoods that border oil refineries, chemical plants, and other such facilities in order to engage residents in conversations about chemical releases from nearby facilities and possible related health issues. The specific goal is to provide residents with information about the iWitness pollution map, including methods of reporting (phone, e-mail, etc.). Students encourage residents to save the reporting number in their phone and they discuss with neighbors how to specifically report possible incidents. They discuss the detailed information that can be included in a report (e.g., specific odors and how to associate a specific odor with a substance) so that this information will be more useful to those who compile the reports and maintain the map.

A regular concern in reporting incidents to local authorities (e.g., the Louisiana Department of Environmental Quality) and requesting that possible releases be investigated is that the response is generally not immediate and odors may have dissipated before an investigation can take place. The map and the ease of reporting have allowed for a permanent record of possible releases to be created.

Students are often skeptical about their ability to have any positive effect in the neighborhoods they visit and they will often comment in their reflective writings that they didn't feel that they were successful in their work. However, as an example of what they can accomplish, a recent visit to a neighborhood in the Algiers section of New Orleans resulted in 349 contacts (out of 655 houses visited), with 76 residents saving the reporting number to the cellphones. The particular neighborhood targeted sits along the banks of the Mississippi River, directly across from a large oil refinery, as well as other smaller facilities.

Over the past few years this type of canvassing work has been completed by students in the class in a number of neighborhoods in the New Orleans area, as well as in Baton Rouge. One neighborhood of particular interest is the Istrouma neighborhood in Baton Rouge. The neighborhood is separated from the ExxonMobil facility, the second largest oil refinery in Louisiana (11th largest in the world), by a raised freeway, with homes sitting just a few short blocks from the near edges of the facility. It is a predominantly African-American neighborhood (90%), with an average household income well below state and national averages. Some 50% of residents have less than a high school education and 41.5% are living below poverty level (*19*).

To get a sense of the concerns some have for those living so close to such a large facility, consider the incident that occurred near the Istrouma neighborhood on June 14, 2012. According to a report submitted to the Louisiana Department of Environmental Quality, as reported in Louisiana Weekly (*20*), Exxon stated that the following amounts of chemicals were released on that date: 28,688 pounds of benzene, 10,882 pounds of toluene, 1,100 pounds of cyclohexane, 1,564 pounds of hexane and 12,605 pounds of other volatile organic compounds. According to Louisiana Weekly article (*20*), Exxon later admitted that the releases were even greater than this.

As also reported by Louisiana Weekly, a later incident at the same facility resulted in release of up to 24 tons of sulfur dioxide a day in May 2013. When interviewed by representatives from the Louisiana Bucket Brigade, significant numbers of residents in the nearby Standard Heights neighborhood reported adverse health effects that they attributed to the SO_2 release.

Subsequent to the work that students conduct in neighborhoods, several assignments are given that ask students to reflect on their service. Students are asked to address the following in a formal paper:

- Do you think that the work you completed will make a difference in the neighborhood? Explain why you feel this way.
- What could or should be done differently if we are to conduct a similar activity in another neighborhood in the future?
- What follow-up work, if any, should be pursued in the neighborhood that we visited?

It is through these papers that the keen interest that students have in this work becomes evident. Despite their earlier skepticism and willingness to be critical of the service activity that was completed, they are very forthcoming in their ideas about future work and it is clear that most have developed a genuine interest in continuing this type of work.

A second assignment is a more informal blog post in which students are asked to share the story of anyone they met while completing the service activity. Excerpts from some of these posts (*12*) related to the visit to Algiers include:

- "At one point this group of ladies we came across sitting outside on their porch started asking us questions about the environment in general and what health effects could be possible from the emissions of the industrial companies. Once we started informing them of the possible emission types they started talking about the sickness they and their grandchildren were feeling around the time of a known industrial mishap."
- "Also, there was an old woman that said she lived in her house for a few decades. She said that her house was getting damaged by the chemicals released by the plants. Her house looked rusted, and she asked for assurance that by reporting the flaring she sees from the factory, that the bucket brigade would get someone to investigate the situation if there were enough reports in the area. She was a little aggressive in asking for assurance, but we said we could make no promises."
- "If a major accident should occur and reach Algiers, it would be a major struggle for the lady to evacuate her house even for a few hours. By the time she could call someone to come and get her and help her leave the chemicals would already be in her system."

In reflecting on work that was completed in the Istrouma neighborhood in Baton Rouge, two students wrote:

- "Ms. Johnson had the most to say (about) the refinery and chemical plants. Ms. Johnson wanted a buyout of the neighborhood she has been living in the area for 23 years and she feels that conditions are only getting worse. She was also animated about the lack of support of her neighbors due to fear of a conflict of interest. There is a lot of people in the neighborhood that puts food on the table by working in one of the plants in the area so it was easy to run into someone who would not speak and give their opinion of the situation. There was a great fear instilled in

- the people of Baton Rouge Istrouma region. There is fear that the people who speak out against the cooperations (*sic*) they will not get any money just in case there is a buyout that takes place in the region."
- "I do not remember his name but I do remember that he had a seventh grade education and his signature was printed. It was precious. Anyhow, everytime I would ask him questions about the area he would give me the same answer, he said that he wasn't trying to put anyone out of a job. He believed that Louisiana Bucket Brigade only wanted to shut down all the plants. We tried to explain that we were just trying to make the air and life cleaner and better for everyone but he was set in his ways. He told me that sometimes they smelled things that smelled bad but the plant was far enough away not to affect them. He believed that the plant was doing their best to keep them safe. I was just sad because he put all his trust in the companies but we know that the companies are not trying their hardest to keep the people of Istrouma safe."

With rare exception, students write very detailed accounts of their encounters. Their stories range from comical and curious to heart-wrenching, but it is abundantly evident that the students feel a strong connection to the residents of the neighborhood they visit. There can be no doubt that many students are truly moved by this experience and have a continued desire to help.

Semester Wrap-up

The culminating assignment for the semester is for students to work in teams of three to deliver an oral presentation summarizing any environmental justice case study of their choosing. Specific areas they cover in these presentations include:

- Details of the environmental hazard, including specific chemicals and specific health hazards associated with these chemicals
- Demographic data for residential areas affected, including race, income, education level, and poverty level
- Specific health issues that have arisen that may be related to the incident or facility
- The responses of both industry and government (local, state, or national) to the situation
- Any legal battles that have ensued
- Recent developments and/or outcomes

Case studies presented can be from anywhere worldwide, so we get to hear about a broad variety of situations. Recent topics have ranged from construction of the Three Gorges Dam in China to problems associated with multiple industrial facilities in East St. Louis, IL. Through these presentations, students exhibit a remarkably better understanding of the concept of environmental justice in comparison to early in the semester. Particularly pleasing is their willingness to research the safety issues and health hazards associated with specific substances mentioned in their case study. The students are normally quite good at critically evaluating the situation and their increased ability to differentiate conjecture from fact is evident.

Toward the end of the semester we always spend some time looking at situations outside of the United States. Of particular interest is the case of Ogoniland in Nigeria, and a number of news reports (*21*) and opinion pieces (*22*) related to this situation are assigned for reading. It is not uncommon to have students from Nigeria or of Nigerian descent in the class, so this particular topic is of great interest to them. For all students this particular case clearly illustrates how devastating the effects of poor or lax regulation can be, but it also allows for a spirited debate regarding just what level of government regulation is appropriate. Problems with spills have been ongoing since the discovery of oil in this area of the Niger Delta in the 1950s and conflicts, sometime violent, involving citizens, government officials, and oil companies continue to this day.

Student Response and Outcomes

Student response to the course has been mostly positive and the learning that takes place is significant, especially for a one-credit course. On course evaluations the majority of students regularly report that opportunities for learning in the course are "Good" or "Excellent". One a scale of poor (1) to excellent (5), the average score on the two most recent semester evaluations was 4.2/5.0.

In these evaluations students often mention that the course helps them to look at regional and world issues in relation to the mission of Xavier. The most common positive comments relate to the opportunities to work with an organization outside of campus and to interact with residents in local neighborhoods. Some specific comments from recent evaluations include:

- "The service-learning project impacted me greatly and made me feel like I've made a difference."
- "The students were able to learn the importance of environmental justice and were able to go into the community."
- "This course allows students to be active in the community and also advocates the concern of world issues as they relate to Xavier's mission."

Despite initial reservations, students generally report that they find the service activities rewarding.

One drawback is that students are currently not allowed to choose their own section of Freshman Seminar, so several each year will have a negative initial reaction to the course, presenting a barrier that is difficult to overcome. In a broad sense, though, students report that they feel equally or more interested in environmental justice (and environmental issues, in general) than before enrolling in the course.

Some students have continued to do volunteer work with the Louisiana Bucket Brigade beyond the scope of the course. Most commonly, they will volunteer to help with the annual New Orleans Earth Day Festival, a sizable event for which the

LABB is the primary organizer. This same event is one in which representatives from the Louisiana Local Section of the ACS and the Xavier Student ACS Chapter regularly participate.

With the ever-changing landscape of environmental justice, the course continues to be dynamic. Issues continue to arise in Louisiana and beyond, and each new group of freshmen brings fresh ideas to the classroom. Most rewarding is that this experience continues to provide opportunities for the instructor to learn alongside eager students.

Acknowledgments

The success of the course is due in large part to the efforts of several people associated with the Louisiana Bucket Brigade. Founding Director Anne Rolfes and Program Manager Anna Hrybyk work with us each year, and past Volunteer Coordinators have included Amelia Rhodewalt, Selena Poznak, and Mariko Toyoji. In Xavier's Center for Student Leadership and Training, Typhanie Jasper-Butler and Kendra Warren have been instrumental in organizing the logistics of our community fieldwork. In the Xavier Center for the Advancement of Teaching, Ross Louis and Mark Gstohl have functioned as Faculty in Residence for Service Learning and they, along with Director Elizabeth Hammer, have provided valuable input in the development of this course. Finally, the 150+ students who have participated in this effort over the past six years deserve thanks for their patience and cooperation as the course has continued to evolve.

References

1. *Environmental Justice Program and Civil Rights*; http://www.epa.gov/region1/ej/ (accessed July 28, 2014).
2. *What Is Environmental Justice?* http://eelink.net/EJ/whatis.html (accessed July 28, 2014).
3. Robinson, D. *Echoes, World Council of Churches*, Issue 17, 2000; http://www.wcc-coe.org/wcc/what/jpc/echoes/echoes-17-02.html (accessed July 28, 2014).
4. Martin, D. S. *CNN Health*, Feb. 26, 2010; http://www.cnn.com/2010/HEALTH/02/26/toxic.town.mossville.epa/ (accessed July 28, 2014).
5. Murphy, T. *Mother Jones*, March 27, 2014; http://www.motherjones.com/environment/2014/03/sasol-mossville-louisiana (accessed July 28, 2014).
6. Arceneaux, C. *Sasol expansion project: What is it?* http://www.kplctv.com/story/24066886/sasol-expansion-project-what-is-it (accessed July 28, 2014).
7. Almasi, D. "Elitist 'Environmental Justice' Bad News for Minorities", *Project 21 New Visions Commentary Editorials*, April 2004; https://www.nationalcenter.org/P21NVAlmasiGreens404.html (accessed July 28, 2014).
8. *Louisiana Bucket Brigade Home Page*; www.labucketbrigade.org (accessed July 28, 2014).

9. *Cancer Alley*; http://en.wikipedia.org/wiki/Cancer_Alley and references therein (accessed July 28, 2014).
10. *iWitness Pollution Map*; http://map.labucketbrigade.org/ (accessed July 28, 2014).
11. *Louisiana Bucket Brigade Refinery Accident Database*; http://www.louisianarefineryaccidentdatabase.org/ (accessed July 28, 2014).
12. *Dr. A's FRSM 100 Weblog*; http://frsm1000dra.wordpress.com/ (accessed July 28, 2014).
13. Lyne, J., "Shintech Picks Louisiana for $1-Billion PVC Complex", *The Site Selection Online Insider*, Jan. 31, 2005, http://www.siteselection.com/ssinsider/snapshot/sf050131.htm (accessed July 28, 2014).
14. Barnett, K. "Motiva Slapped with $500,000 Fine". *The St. Charles Herald Guide*; Oct. 26, 2012, http://www.heraldguide.com/details.php?id=11572 (accessed July 28, 2014).
15. Whitehouse Home Page. "*Obama Administration Convenes Environmental Leaders at Historic White House Environmental Justice Forum Featuring Five Cabinet Secretaries*"; White House Council on Environmental Quality press release, Dec. 15, 2010 retrieved from http://www.whitehouse.gov/administration/eop/ceq/Press_Releases/December_15_2010 (accessed July 28, 2014).
16. Sutley, N., Perciasepe, B. "*Real Progress on Environmental Justice*"; Council on Environmental Quality, March 20, 2013 retrieved from http://www.whitehouse.gov/blog/2013/03/20/real-progress-environmental-justice-0 (accessed July 28, 2014).
17. Wallace, P. E. Environmental Justice and the BP Oil Spill: Does Anyone Care About the "Small People" of Color? *The Modern American*; Washington College of Law, Vol.6, Issue 2, 2010, retrieved from http://www.wcl.american.edu/environment/WallaceVolume6Issue2.pdf (accessed July 28, 2014).
18. Whitehouse Home Page. "*Executive Order: Improving Chemical Facility Safety and Security (Executive Order 13650*" retrieved from http://www.whitehouse.gov/the-press-office/2013/08/01/executive-order-improving-chemical-facility-safety-and-security (accessed July 28, 2014).
19. Data retrieved from *Istrouma neighborhood in Baton Rouge, Louisiana (LA), 70802, 70805 detailed profile*; http://www.city-data.com/neighborhood/Istrouma-Baton-Rouge-LA.html (accessed July 28, 2014).
20. Buchanan, S. "ExxonMobil is Scrutinized in Baton Rouge After Past Leaks". *Louisiana Weekly*, July 15, 2013 retrieved from http://www.louisianaweekly.com/exxonmobil-is-scrutinized-in-baton-rouge-after-past-leaks/ (accessed July 28, 2014).
21. Watkins, T. "*Amnesty Accuses Shell of Making False Claims on Niger Delta Oil Spill*". CNN, November 7, 2013, retrieved from http://www.cnn.com/2013/11/06/world/africa/nigeria-shell/ (accessed July 28, 2014).
22. Jethoha "*Oil, Blood, and Fire: Environmental Assessment of Ogoniland, Nigeria*" retrieved from http://mastereia.wordpress.com/2013/10/08/oil-blood-and-fire-environmental-assessment-of-ogoniland-nigeria/ and references therein (accessed July 28, 2014).

Chapter 3

Green Action Through Education: A Model for Fostering Positive Attitudes about STEM by Bringing the Scientists to the Students, Not the Other Way Around

William J. Donovan,*,[1] Ethel R. Wheland,[2] Gregory A. Smith,[3] and Angela Bilia[4]

[1]Department of Chemistry, The University of Akron, Akron, Ohio 44325-3601
[2]Department of Mathematics, The University of Akron, Akron, Ohio 44325-4002
[3]Department of Biological Sciences, Kent State University at Stark, 6000 Frank Avenue NW, North Canton, Ohio 44720-7599
[4]Department of English, The University of Akron Akron, Ohio 44325-1906
*E-mail: wdonovan@uakron.edu.

>Our chapter describes an integrated, interdisciplinary thematic learning community based on the SENCER model to draw students with less-than-optimal indicators of college success to STEM majors. We describe the learning community, the interdisciplinary model, and the results of our project.

A goal of science education should be to show students how science functions, regardless of whether they will be following science careers (*1*). The SENCER (Science Education for New Civic Engagements and Responsibilities) model is designed to improve undergraduate STEM learning. However, most SENCER courses are standalone courses in a science field, often for Honors or upper division students (*2*). The model described in this chapter is an academic design that allows first-year students whose indicators suggest less-than-optimal STEM preparation to engage in STEM content within the context of a civic issue.

© 2014 American Chemical Society

As reported in *Rising Above the Gathering Storm*, "introductory science courses can function as 'gatekeepers' that intentionally foster competition and encourage the best students to continue, but in so doing they also can discourage highly qualified students who could succeed if they were given enough support in the early days of their undergraduate experience" (*3*). The model in this paper was organized to pique undergraduates' interest in science, remove or diminish the competitive atmosphere surrounding science classes, and show students of all abilities how science functions. It is unique in its design to have scientists bring their STEM expertise to students in non-traditional course settings rather than having students meet the scientists in a traditional university science course.

Addressing the challenges of STEM reform can be viewed on three levels: high-school science and math instruction, limitations of current reform strategies and obstacles caused by disciplinary silos. Contact with K-12 teachers suggests that emphasis on graduation tests and uniform science and math curricular standards in high school have led to a reduction in inquiry and critical thinking skills, yet we as college STEM faculty believe that students possess these skills, which are needed for success in STEM courses. Our project aimed to incorporate effective teaching pedagogies involving learning communities, the SENCER model, and student engagement, in order to narrow the K-12/College gap and prepare students for STEM majors while enhancing science literacy. The long term challenge is to transcend course boundaries in order to promote campus-wide scientific literacy, meeting NSF's challenge for effective science education for all. What is needed is a mechanism that allows students the opportunity to engage STEM content within the context of a civic issue and within their discipline. Effective mechanisms allow students to engage in a high impact activity during their first year and then an activity "… taken later in the major field with common intellectual content so many students across the campus can engage and discuss the same material both in and outside class" (*4*).

We at the University of Akron (UA) wrote and were funded for a NSF CCLI Phase II grant with partners at Indiana University-Purdue University Indianapolis (IUPUI) and Lorain County Community College. UA and IUPUI have years of experience with learning communities (LCs), drawing national attention for their efforts in creating an exemplary first-year experience. UA has also had considerable success in using research-based evidence to develop highly-organized campus-wide curricular reforms. Annually, our *Make a Difference Day* service learning project attracts 1000 participants. We are especially proud of our unique *Rethinking Race: Black, White and Beyond* program, an intense campus-wide series of events focused on diversity education. More than 60 courses integrate *Rethinking Race* topics via external keynote speakers, face-to-face discussion sessions and a myriad of Student Affairs events which impacted more than 3300 students representing more than 100 majors. Among other things, we learned that a small committee of faculty champions can develop a program that reaches many and is embraced by students and faculty. Faculty welcomed the opportunity to integrate the "big question" with their own course material. *Rethinking Race* transformed a commuter-campus while breaking down disciplinary silos. The STEM speaker series, initially titled *Rethinking STEM,* used the same successful strategies and assessments as the *Rethinking Race* model. We used our expertise to show that non-science majors can be energized by a scientific civic/humanitarian issue when the issue is seamlessly designed into *their* major courses.

UA is a public-assisted, research-intensive metropolitan university enrolling 26,000 students, 15% from underrepresented groups and 10% of whom are residential, with a first-time, first-year entering class of 4,000 students. UA has earned a number of Carnegie Foundation accolades including both *Curricular Engagement and Outreach & Partnerships* classifications and leadership in the *Critical Thinking for Civic Thinking* CASTL program. UA has actively participated in many initiatives including summer STEM academies for K-12 students, K-12 teacher education workshops, and the recent 5-year Choose Ohio First STEM scholarship program similar to NSF's S-STEM program. UA has 57 LCs, including 4 exploratory LCs dedicated to students who are undecided about a major and career. UA's mission is to be differentiated as a STEM institution by increasing the number of STEM majors which is aligned with strategic plans of the University System of Ohio.

The Overall Project and Goals

To implement this project, students and staff pursued activities using an interdisciplinary curriculum built around investigating the civic issue of water quality rights and responsibilities. In addition to increasing the students' understanding about the civic issue, the project had a related objective of increasing students' interest in the pursuit of majors in the STEM fields of science, technology, engineering, and mathematics. The project aimed to implement the innovative NSF-funded SENCER model, build faculty expertise and sense of community, and create new teaching materials and strategies. The overarching goal was to enhance student scientific literacy by thematic integration of curricula. There were four major goals of the project:

Goal 1: *Implement Educational Innovation.*

Implement best practices as espoused by the NSF CCLI project SENCER in unique ways by creating flexible educational strategies that provide undergraduate students on each of our campuses the opportunity to engage in civic issues and the underlying STEM concept and applications.

Goal 2: *Impact on Students*

Create an environment that will enhance the development of civic engagement, critical thinking skills, reduce student anxiety, enhance student attitudes and self-efficacy, improve science and math ability, and increase awareness of the academic preparation needed for STEM careers.

Goal 3: *Build STEM Faculty Community*

Design and assess the processes for an effective SENCERized Learning Community model.

Goal 4: *Offer STEM for All*

Design and assess the innovative and novel educational model "*Rethinking STEM: Science beyond the Classroom*" that integrates a STEM-driven civic issue into multiple disciplines across the university to enhance science literacy as well as faculty and student engagement.

The Model

This model was patterned after a standard learning community structure for its intentional linkage of courses and coursework (5). In 2009 and 2010, the LC included 11 credit hours as a fully-integrated four-course block consisting of English Composition, Oral Communication, Student Success Seminar, and Career Planning (Table 1). In 2012, due to the redesign of the speech course, it was discontinued from the LC. To address the issue of students receiving general education science credit for the significant STEM content of the special topics course, an existing 3-credit course (Environmental Science & Engineering) was used so that students would receive such credit should they not choose to pursue a STEM major, so that they would get some benefit in their progress toward graduation. These changes, along with English Composition being reduced to a 3-credit-hour course university-wide in 2012, made the LC block total 6 credit hours (Table 2). Students also enrolled in an appropriate Mathematics course based on placement scores. As well as allowing time for outdoor and classroom laboratory exercises, the block schedule provided a flexible calendar to integrate the civic theme with science applications across disciplines. Intentionally omitted from the schedule was a formal, traditionally-structured science class. Instead, STEM scientists coordinated both indoor and outdoor labs with the students.

The goal was to create an engaging collaborative environment that would foster development of critical thinking skills, reduce student anxiety, enhance student attitudes and self-efficacy, improve science and math ability, and increase awareness of the academic preparation needed for STEM careers—outside of the parameters of the traditional university science course. At different institutions, teams of faculty could tailor such a model to their situations. The English and Speech classes were back to back in order to provide a two-hour morning block which was occasionally used for activities such as field sample collection.

Because the civic issue was the conduit through which science concepts were introduced to students, it had to be chosen carefully. It needed to invite students "to put scientific knowledge and scientific method to immediate use on matters of immediate interest," and to involve a contested issue (6). Adopting water quality as the civic issue seemed logical given that the setting for this project is in a city engaged in a legal battle with the U.S. Environmental Protection Agency (EPA) over a combined sewer system that discharges into the Cuyahoga River. This river was immortalized in songs by Randy Newman ("Burn On", 1972), R.E.M. ("Cuyahoga", 1986) and Adam Again ("River on Fire", 1992) after an infamous 1969 fire (7–9) that precipitated the environmental movement.

Table 1. LC Block Schedule (Fall Semester 2009 and 2010)

	Mon	*Tues*	*Weds*	*Thurs*	*Fri*
8:50-9:40	English	English	English		English
9:55-10:45	Speech		Speech		Speech
11:00-11:50					
12:05-12:55	Math	Math	Math		Math
1:10-2:00		Student Success Seminar		Student Success Seminar	
2:15-5:00				Career Planning /Special Topics	

Table 2. LC Block Schedule (Fall Semester 2012)

	Mon	*Tues*	*Weds*	*Thurs*	*Fri*
9:55-10:45	English	English	English		English
11:00-11:50					
12:05-12:55	Math	Math	Math		Math
1:10-2:50		Envi. Sci. & Eng.		Envi. Sci. & Eng.	
2:50-3:50					

This uniquely structured learning community, called GATE (*Green Action Through Education*), was designed to immerse students in the sciences in a more natural way than through a formal science course. Our STEM-related goals were to have students react actively rather than passively to multiple aspects of the selected civic issue, to communicate effectively, and to understand and know how to work in a laboratory environment, be it indoors or outdoors. The LC STEM-specific learning objectives and indicators are listed in Table 3.

Table 3. LC STEM-Specific Learning Objectives and Indicators

At the completion of the 15-week GATE Learning Community, students should demonstrate:
1. Engagement in multiple aspects of a civic issue.
   (a) Develop abstract and concrete scientific thinking skills and awareness of the "big picture"
   (b) Discuss how people and systems affect the watershed and thus water quality
2. Ability to communicate effectively, both orally and in writing.
   (a) Use appropriate terminology in communicating concepts (b) Write a lab report
   (c) Use simple graphs, tables, and charts to communicate information
3. Ability to work effectively and safely in a laboratory environment, whether the lab is a classroom or an outdoor setting.
   (a) Work in teams as well as independently
   (b) Observe laboratory safety and housekeeping protocol
   (c) Record experimental work in lab notebooks
   (d) Perform accurate quantitative measurements, interpret experimental data and results, perform calculations on results
   (e) Demonstrate increased skill in using algorithms and in doing calculations, measured numbers vs. exact numbers, significant figures, dimensional analysis, and proportionality
   (f) Apply math/statistical concepts such as extrapolation, approximation, statistical validity to solve scientific problems
   (g) Report data honestly and ethically

Our approach to integrating science aspects of the civic issue into the LC in a non-traditional way demanded non-traditional approaches to meeting our learning objectives. The scientists from Chemistry, Biology, Math, and Public Health challenged and engaged the students in activities with scientific underpinnings that demanded collaboration and participation. To integrate the theme across the LC, the English, Speech, and Career instructors guided students in writing about, talking about, debating about, and considering career plans related to water quality in juxtaposition with the presentations of STEM aspects of water quality.

In order to get students to think about how people and systems and they themselves affect water quality, the scientists asked students to maintain a water log after having them tour the sewage treatment plant. At the plant, staff talked about the biological and chemical aspects of sewage treatment for a city with an older combined sewage system. The odors and aromas, sights, and sounds of huge quantities of water flowing into the plant for treatment and then being released into the river gave the students a sensory experience that could not be duplicated by mere descriptions and words. Capitalizing on students' budding awareness that humans do make an impact on water quality, the scientists asked students to maintain a detailed water log for 4 days to document their own contributions to this impact. The analysis of the water log was the first of many occasions for the scientists to address the math skills covered by Learning Objectives 3d, 3e, and 3f.

The scientists also introduced the students to ways of directly measuring water quality by taking them to sites on the river upstream and downstream from the sewage treatment plant outfall to collect water samples and then to the chemistry lab to analyze the samples for ammonium nitrogen, nitrate nitrogen, calcium ion, chloride ion, total dissolved solids, and pH (see Appendix 1).

To introduce these students to the sound, but not so well-known, notion that water quality can be measured by indirect methods, the scientists twice brought the students to a local stream to collect and count fish and macroinvertebrates, healthy populations of which are good indicators of high water quality. The students were able to work with an employee of the Ohio EPA, a world leader in developing biotic indices for terrestrial and aquatic habitats. The students also field-tested a new sampling method being developed by the EPA for the collection of stream quality data by volunteers. Based on individual feedback, some of the most memorable experiences for the students were electroshocking fish and manually sampling for aquatic insects and crustaceans. We were not required to deal with permitting issues because the activities were covered by the EPA's already-existing permits.

To provide models for drawing scientifically informed conclusions about water quality, the scientists worked with students to create Excel graphs of *E. coli* populations measured in the river and discuss the effects of their findings on the ways they would use the river. Students also visited current and former dam sites and observed a debate between experts about whether to maintain or destroy dams in the future so that they could listen for the ways in which the experts had formed their scientifically informed conclusions. Table 4 lists the above engagement activities and learning objectives underlying those experiences.

Table 4. LC Engagements and Underlying STEM-Specific Learning Objectives

Engagement	*Underlying Learning Objective (detailed in Table 3)*
Treatment plant visit	1a, 1b
Two Yellow Creek trips	1a, 1b, 3a, 3b, 3c
Water Log calculations/math	1a, 1b, 3d, 3e, 3f, 3g
Dam visits/debates	1a, 1b, 3b, 3c, 3d, 3g
Excel activity	1a, 1b, 2c, 3d, 3f
Chemistry lab analysis	1a, 1b, 2a, 2b, 2c, 3a, 3b, 3c, 3d, 3g

Findings about Student Attitudes and Habits of Mind

GATE students completed an Attitudes Towards Science Inventory (ATSI), adapted from the TOSRA (*10*, *11*), to assess their attitudes towards various science activities and concepts. Table 5 shows a summary of the ATSI findings for the 2010 GATE LC and a comparison LC. The ATSI used scaled-response items on a scale of 1 (strongly disagree) to 6 (strongly agree), so the range shown in Table 5 indicates the lowest possible subtotal (all ratings of 1) to highest possible (all 6) subtotal for each category. In each significant finding, the GATE LC scored higher than the comparison LC.

Table 5. All Significant (p<.01) ATSI Pre/Post Findings for GATE and Comparison LCs, Fall 2010

ATSI sub-category	Range	GATE LC			Comparison LC		
		N	Pre	Post	N	Pre	Post
Science Self-Concept	9-54	15	38.20	40.33	27	31.78	31.59
Science in Daily Life and Future	6-36	15	30.33	29.73	26	18.38	18.27
Attitudes towards Science Education	3-18	16	15.37	15.50	28	11.57	10.82
Comfort with Doing Science Activities	4-24	16	21.31	20.38	28	17.43	16.96
Science in the Real World	6-36	15	27.07	27.80	27	24.44	24.59

In addition, the 2010 GATE LC students completed a self-assessment of Habits of Mind (*12, 13*). A staff member also evaluated students to observe the frequency with which they demonstrated one or more of the Habits of Mind attributes. Table 6 provides a brief summary of the findings showing the four Habits of Mind where significant increases in both student self-rating and staff rating were found. The different descriptors used in the two rating rubrics might explain why students rated themselves higher than the staff member rated them.

Table 6. Student and Staff Ratings of Habits of Mind Showing Significant Changes, Fall 2010

Habit of Mind	GATE Student Self Ratings[a]			Staff Ratings of Students[b]		
	Pre	Post	t	Pre	Post	t
Awareness of own thinking	1.82	2.35	4.21[c]	1.62	1.87	2.23[d]
Accuracy and search for accuracy	1.81	2.40	5.83[c]	1.50	1.75	2.23[d]
Clarity and search for clarity	2.11	2.35	2.70[d]	1.50	1.75	2.23[d]
Demonstration of pushing the limits of own knowledge and ability	1.43	2.00	3.57[c]	1.12	1.37	2.23[d]

[a] Scale: 0 = *Not at all* through 3 = *A lot* [b] Scale: 0 = *Least accomplished* through 3 = *Most accomplished* [c] p<.01 [d] p<.05

Discussion

In immersing students in science via a civic issue, the civic issue should be one with local importance and one that matters to the student participants. In Kansas, for instance, it would make no sense to talk about Ohio's Cuyahoga River as the primary "hook." This excerpt from one student's portfolio demonstrates her engagement and involvement with the civic issue.

- *For the lab where we had to test drinking water, I tested two samples of the well water. One was only a day old out of the distillers and the other was two months sitting in a jug. I was curious to see if there were any differences for health issues. Jugs of water can sit in our boat for up to at least a month and a half. After tests, the results showed a very slight difference. All the ions showed a range of only 0-1.5 mg/L. The total dissolved solids were between 0-7. The only slight difference was the average pH for the jugged water was higher by 0.3, but even that is not bad because the pH measurements ranged from 6.78-7.25. While discussing the lab with the class, I realized the well water my family and I drink is one of the cleanest in the room.* (2010 portfolio reflection)

Another student wrote:

- *The learning community has made me realize how important science is to the world; not only discovering new technologies, but also everyday things like water quality. . . . I think a large part of why the learning community opened my eyes was because of the variety of ways science was taught.* (2010 portfolio reflection)

The civic issue for the GATE LC in 2009 was broadly stated as "Water quality in the Cuyahoga valley." In the 2010 LC, we asked the students to address the more specific question "What responsibilities do local municipalities have in maintaining quality recreational water?" In comparing the student evaluations for the two years, we learned that the 2009 topic worked better than the 2010 topic, because the 2009 topic allowed students to discover direct relationships between their consumption of water and the civic issue. Another lesson learned as the scientists brought their STEM expertise to the students was that that they needed to offer more step-by-step instructions to non-science majors when asking them to demonstrate the indicators of the STEM-specific learning objectives. When the scientists and students worked together in the lab during the 2009 LC, it became obvious that the scientists had overestimated the students' comfort level and ability in the chemistry lab. The following year's chemistry lab activities were greatly revised to provide more step-by-step guidance.

One of the aspects of the 2010 topic that immediately engaged students and also contributed to a change in water consumption was the monitoring of their own water usage via a water log (see Appendix 2). The following are students' comments about that activity.

- *The water log activity we did for instance, made me realize how much water I use at a time. After that activity I realized how much water was used by the six billion people in the world and my mind was blown. The positive side of realizing how wasteful people are is that we can change and try and fix everything we have done.* (2010 portfolio reflection)
- *By doing the water log and figuring out how many gallons of water was used in only four days was very eye opening. This helped me to change my water usage amount. I never realized how much water I waste daily. I have changed how long of showers I take because this is the biggest water waster for me.* (2010 portfolio reflection)

The water log activity continued to be used in the 2012 LC as the capstone project, worth about a quarter of the total points in the 3-credit engineering course. The activity has been combined with the water quality analysis work in the general education chemistry lab course for non-STEM majors, Chemistry 101, in a two-part lab activity dealing with both water quality and the math and economics of water usage, to bring the civic issue into that larger class. The water log activity used in the general education course to this day is shown in Appendix 3.

This integrated program, designed to bring engaging science to students, would not have been successful without the efforts of the instructors of non-STEM courses. The English course played a huge role in the integration of the message *(14)*. As students wrote about their experiences with science in these non-traditional settings, the challenges of writing presented themselves. When students were faced with writing what they felt were very challenging and sophisticated scientific papers, their use of correct grammar, spelling, sentence structure, sentence variety, and other fundamental elements of writing suffered. While these issues could be addressed in subsequent revisions, the students' occasional lack of even the most basic scientific knowledge sometimes slowed the revision process, as did their inability to grasp a variety of scientific concepts in one semester. While it wasn't easy, the significant effort on the part of the English professor and the students resulted in students responding well and learning much from the constant revision of their work.

The positive outcome of having STEM and non-STEM faculty work together to bridge some of the gaps between their disciplines was noticed and commented upon by students. As one student observed:

- *Having the [instructors] in ... different areas of college like general courses working together ... has taught us how to write a science-based paper... [and] has taught us how to speak and how to make a speech for a science audience.... It's not traditional [compared to] what all my other friends are doing in their freshman classes.... Having the faculty that don't have the science background helps give you that wide array of skills.* (2009 interview)

Conclusion

Our model of having scientists work with students in diverse settings proved to be very effective in fostering favorable attitudes about STEM. Students' reactions to actually working with scientists in the field were uniformly positive. The water quality issue worked well to introduce these non-STEM majors to meaningful science. Students who started the semester with a somewhat passive response to water quality issues became much more involved in monitoring their water usage and in studying the quality of the water they drank. They indicated that they were better-equipped to discuss this issue scientifically than in the past.

On the other hand, the project was labor-intensive for all involved. For instance, the struggles of non-STEM students to write and speak about what was going on in the scientific realm created an additional burden for their instructors. These struggles revealed to the scientists in the group that some of their expectations of the readiness of college freshmen to engage fully in science classes might also need to be adjusted. The project evaluators asked both STEM and non-STEM faculty if they had to change their teaching strategies to accommodate the goals of the project. Both STEM and non-STEM professors indicated that two-thirds to three-quarters of their class writing assignments were about their field trips and lab work. A STEM professor said that lab work had to be organized differently because the STEM LC students did not have the required background in a lab setting as compared to typical students in science courses. Many STEM LC students did not understand the need for rigor and patience with scientific activities during the pre-experimental and experimental stages.

In addition, career-planning course activities also changed because of the need to be flexible and responsive to the science content that needed to be integrated into activities. Another adjustment was that the order of course content was changed to make courses more meaningful to what the students were doing in science. Finally, there was always a focus on creating a non-threatening environment for students.

The faculty had mixed feelings regarding the likeliness of students to be engaged in the civic issue after the LC experience concluded. Some faculty felt that the STEM LC students would be less likely to be engaged in the civic issue because of the lack of opportunity, preoccupation with their academic focus, and because faculty had not guided or advised students along those lines. This identified gap in guidance provided part of the rationale for the revision of the project plan in the second year. Faculty also felt that the experiences in the LC were so unique compared to traditional college classes that students would remember the experience for a long time and could seek to pursue their interests in the civic issue.

Several GATE LC faculty observed that while for some students the problem solving and critical thinking were observed within the content of their discussions, other students seemed to be learning for the first time that science and technology had such an impact on the quality of their lives. One student was overheard saying, "every time I have a bottle of water, a glass of water, water from a drinking fountain – however I get drinking water – I think about what is the content in that container, what am I ingesting." Many students had proclaimed by the end of the semester that they would never think about water in the same way. Other students demonstrated a lack of development in the habits of the mind such as openness to new ideas, accepting alternative solutions, or deciding what to believe.

Acknowledgments

This material is based upon work supported by the National Science Foundation under CCLI Phase II Grant No. 0920453.

The authors wish to acknowledge UA colleagues Todd Wagler (Chemistry Lab Supervisor), Annabelle Foos (Geology), Tom Dukes (English), Justin Brantner (Education), Bob Crowley (Communications), Shane Strnad (Nursing), Kathy Ross-Alaolmolki (Nursing), Jennifer Hodges (University College), and Bonnie Williams (University College). We also thank external evaluators Deborah Shama-Davis and Lennise Baptiste (Kent State University) and Tom Wood (SENCER) for their support and participation. Finally, we could not have undertaken the proposal or the project itself without the dedication and tireless effort of our late UA colleague Helen Qammar (Chemical Engineering) and dedicate our work to her memory.

References

1. Osborne, J.; Dillon, J. *Science education in Europe: Critical reflection. A report to the Nuffeld Foundation*; King's College London: London, U.K., 2008; p 7.
2. Haines, S. *J. Coll. Sci. Teach.* **2010**, *39* (3), 16–23.
3. National Academy of Sciences. *Rising Above the Gathering Storm: Energizing and Employing America for a Brighter Economic Future*; Committee on Science, Engineering, and Public Policy (COSEPUP); National Academies Press: Washington, DC, 2007; pp 102–110.
4. Kuh, G. *High-impact educational practices: What they are, who has access to them, and why they matter*. Association of American Colleges and Universities and LEAP: Washington, DC, 2008.
5. Smith, B. L. *Lib. Educ.* **1993**, *79* (4), 32–39.
6. *Sencer Ideals*. SENCER Institute; http://www.sencer.net/About/sencerideals.cfm (accessed October 2014).
7. Adler, J. H. *Fordham Environ. Law J.* **2003**, *14*, 89–146.
8. U.S. Environmental Protection Agency (EPA). *Cuyahoga River area of concern*; 2011;http://www.epa.gov/greatlakes/aoc/cuyahoga/ (accessed October 2014).
9. Krugman, P. Drilling, disaster, denial. *The New York Times*; May 2, 2010; http://www.nytimes.com/2010/05/03/opinion/03krugman.html (accessed October 2014).
10. Fraser, B. J.. *TOSRA test of science-related attitudes handbook*; Australian Council for Educational Research Limited: Hawthorn, Victoria, Australia, 1981.

11. *Research Tools @ FSU*; http://ret.fsu.edu/Research_Tools.htm (accessed October 2014).
12. *Discovering and Exploring Habits of Mind*; Costa, A. L., Kallick, B., Eds.; Association for Supervision and Curriculum Development: Alexandria, VA, 1993.
13. Marzano, R.; Pickering, D.; McTighe, J. *Assessing Student Outcomes: Performance Assessment Using the Dimensions of Learning Model*; Association for Supervision and Curriculum Development: Alexandria, VA, 1993.
14. Ghent, C. *J. Coll. Sci. Teach.* **2010**, *39* (3), 34–38.

Appendix 1

Appendix 1

A Chemist's Approach to the Assessment of Water Quality
GATE Learning Community, Fall 2010

Sources:
- 3150:154 lab manual, Dr. Todd Wagler
- *Water Quality with Vernier*

Tuesday, Oct. 5 1:10-2:00 MGH 306	Wednesday, Oct. 6 8:50-10:45 (replacing English & Speech class periods)	Thursday, Oct. 7 1:10-2:25 MGH 306♦ 2:25-5:00 KNCL 304♣
• Dr. Donovan sets stage for chemist's analysis of water • Dr. Donovan discusses lab protocol	• collect samples at various river locations ○ test with pH paper	• ALICE training ♣ chemistry lab analysis of samples collected previous day
Tuesday, Oct. 12 1:10-2:00 MGH 306	**Wednesday, Oct. 13** 7:00-8:00 SU Theatre■ ~8:30pm at home/dorm♠	**Thursday, Oct. 14** 1:10-2:00 SU Theatre■ 2:00-5:00 KNCL 304♣
• Dr. Qammar discusses Excel • distribute Akron Drinking Water Quality Report • students pick up storage bottle for water sample to be collected at home on Wednesday evening	■ suggested: attend Dr. Saskowsky's talk ♠ collect drinking water sample(s)	■ SJA presentation ♣ chemistry lab analysis of drinking water samples collected by students

When thinking of recreational opportunities that are water-related, we might list such things as fishing, boating, and swimming. Each of these activities is dependent on high-quality water. What are some ways that a chemist can determine the quality of the water in which we are fishing, boating, or swimming?

There are various ways to assess water quality. As chemists, we are able to measure the amounts of substances in the water such as calcium, nitrate, ammonium nitrogen, and chloride, as well as the acidity of the water and the total concentration of solids dissolved in the water. Why would we want to do so?

- A. If the **calcium ion concentration** in freshwater drops **too low**, then the water can support only sparse plant and animal life, a condition known as *oligotrophication*.

- B. On the other hand, **high nitrate ion concentrations** contribute to excessive growth of aquatic plants and algae, a condition called *eutrophication*. Unpleasant odor and taste of water, as well as reduced clarity, often accompany this process.

- C. **High ammonium nitrogen** levels in surface waters can be toxic to some aquatic organisms. Even levels that are only moderately high can cause excess plant and algal growth. However, if **too little ammonium nitrogen** is present, plant and algal growth may be curtailed.

- D. **Chloride ion** levels that are **too high** are toxic to many freshwater organisms.

- E. **Acidity** of water as measured by pH is a very important indicator of water quality, because of the sensitivity of aquatic organisms to the pH of their environment. Small changes in pH can endanger many kinds of plants and animals.

- F. The concentration of **total dissolved solids (TDS)** is sometimes used as a "watchdog" environmental test. A high concentration of dissolved solids is not necessarily an indication that a stream is polluted or unhealthy. However, **high TDS levels** may be associated with unpleasant mineral taste or high or low acidity, depending on what specific solids are dissolved.

Appendix 1-1

On Wednesday, October 6, we will collect water from locations upstream and downstream from the sewage treatment plant and chemically analyze it on-site. On Thursday, we will go to the chemistry lab to conduct further experiments and will then determine how it measures up as recreational water.

When you come to the chemistry lab on Thursday, your team will pick up your water sample and start the chemical analysis at one of six stations (A–F) to which you will be assigned. You will rotate through all of the remaining stations to complete the analysis of your water.

You will have 10 minutes at each station. You will follow these same steps at each station to analyze a sample for each of the six important water quality measures.

1. After gently shaking your container, pour a small amount of the water sample into a clean, dry beaker.
2. Place the tip of the sensor into the sample beaker.
3. <u>Hold the probe tip still</u> for at least 10 seconds. Record in the Data Table the reading displayed on the screen.
4. Repeat data collection on a second portion of your water sample, repeating steps 1-3. You should expect your two readings to be close to each other, but not identical. To minimize the effect of readings that vary too much, take a third and a fourth measurement.
5. Either now or as soon as possible after leaving the station as time permits, average your four readings and record the average along with the individual readings.

Repeat these steps for each sample that your group receives.

Appendix 1-2

Sample No. _____ Source _____

Data

Station		A Calcium Ion (Ca^{2+}) mg/L	B Nitrate Ion (NO_3^-) mg/L	C Ammonium Ion (NH_4^-) mg/L	D Chloride Ion (Cl$^-$) mg/L	E pH	F Total Dissolved Solids (TDS) mg/L
Readings	1						
	2						
	3						
	4						
Average of four readings							

Comparison of Our Group's Findings with Expected Levels

	Calcium Ion (Ca^{2+}) mg/L	Nitrate Ion (NO_3^-) mg/L	Ammonium Ion (NH_4^+) mg/L	Chloride Ion (Cl$^-$) mg/L	pH	Total Dissolved Solids (TDS) mg/L
Expected level						
Our measured level						
Difference (+ or −)						
Notes						
Conclusions						

Appendix 1-3

Sample No. _____ Source _____

Data

Station		A Calcium Ion (Ca^{2+}) mg/L	B Nitrate Ion (NO_3^-) mg/L	C Ammonium Ion (NH_4^-) mg/L	D Chloride Ion (Cl^-) mg/L	E pH	F Total Dissolved Solids (TDS) mg/L
Readings	1						
	2						
	3						
	4						
Average of four readings							

Comparison of Our Group's Findings with Expected Levels

	Calcium Ion (Ca^{2+}) mg/L	Nitrate Ion (NO_3^-) mg/L	Ammonium Ion (NH_4^+) mg/L	Chloride Ion (Cl^-) mg/L	pH	Total Dissolved Solids (TDS) mg/L
Expected level						
Our measured level						
Difference (+ or −)						
Notes						
Conclusions						

Appendix 1-4

A. Calcium Ion, Ca^{2+}

Calcium, in the form of the Ca^{2+} ion*, is one of the major positive ions in saltwater and freshwater. Most calcium in surface water comes from streams flowing over limestone, gypsum, and other calcium-containing rocks and minerals. Groundwater and underground aquifers leach even higher concentrations of calcium ions from rocks and soil.

Sources of Calcium Ions
• Limestone: $CaCO_3$
• Dolomite: $CaCO_3$–$MgCO_3$
• Gypsum: $CaSO_4 \cdot 2H_2O$

Expected Levels

The concentration of calcium ions in freshwater is found in a range of 0 to 100 mg/L. A level of 50 mg/L is recommended as the upper limit for drinking water. High calcium ion levels are not considered to be a health concern; however, water with calcium levels above 50 mg/L (commonly called "hard water") can be problematic due to formation of calcium deposits in plumbing or decreased cleansing action of soap. If the calcium ion concentration in freshwater drops below 5 mg/L, it can support only sparse plant and animal life, a condition known as *oligotrophication*.

B. Nitrate Ion, NO_3^-

Nitrate ions found in freshwater samples result from a variety of natural and manmade sources. Nitrates are an important source of nitrogen for plants and animals to synthesize amino acids and proteins. Most nitrogen on earth is found in the atmosphere in the form of nitrogen gas, N_2. Through a process called the *nitrogen cycle*,[†] nitrogen gas is changed into forms that are useable by plants and animals. These conversions include industrial production of fertilizers, as well as natural processes, such as legume-plant nitrogen fixation, plant and animal decomposition, and animal waste.

Sources of Nitrate Ions
• Agriculture runoff
• Urban runoff
• Animal feedlots and barnyards
• Municipal and industrial wastewater
• Automobile and industrial emissions
• Decomposition of plants and animals

Expected Levels

The nitrate level in freshwater is usually found in the range of 0.1 to 4 mg/L. Unpolluted waters generally have nitrate levels below 1 mg/L. The effluent of some sewage treatment plants may have levels in excess of 20 mg/L. Other manmade sources of nitrate may elevate levels above 3 mg/L. These sources include animal feedlots and runoff from fertilized fields. Nitrate levels above 10 mg/L in drinking water can cause Blue Baby Syndrome in infants, a disease where nitrate ions convert hemoglobin into a form that can no longer transport oxygen.

High nitrate concentrations contribute to *eutrophication*, the excessive growth of aquatic plants and algae in lakes and ponds. Unpleasant odor and taste of water, as well as reduced clarity, often accompany this process.

---

* An *ion* is an atom or group of atoms that carries an electrical charge.
† See *Ammonium Ion* on the next page for further information on the nitrogen cycle.

Appendix 1-5

C. Ammonium Ion, NH_4^+

The ammonium ion is an important member of the group of nitrogen-containing compounds that act as nutrients for aquatic plants and algae.

Sources of Ammonium Ions
- Decaying plants and animals
- Animal waste
- Industrial waste effluent
- Agricultural runoff
- Atmospheric nitrogen

All plants and animals (including humans!) require nitrogen as a nutrient to synthesize amino acids and proteins. Most nitrogen on earth is found in the atmosphere in the form of N_2, but plants and animals cannot utilize it in this form. The nitrogen must first be converted into a useable form, such as ammonium ion or nitrate ion. These conversions among the various forms of nitrogen form a complex cycle called the *nitrogen cycle*.

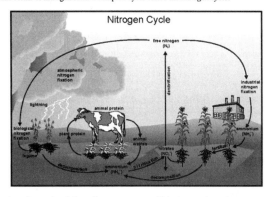

If ammonium nitrogen levels in surface waters are too high, they can be toxic to some aquatic organisms. If the levels are only moderately high, plant and algal growth will usually increase due to the abundance of nitrogen available as a nutrient. Dissolved oxygen levels may also be lowered when ammonium nitrogen is high.

If enough nutrients are present, *eutrophication* may occur. In some cases, this process can become so advanced that the body of water may become a marsh, and eventually fill in completely.

Expected Levels

Ammonium nitrogen levels are usually quite low in moving surface waters because there is little decaying organic matter collecting on the bottom. If there is a high level of ammonium nitrogen in a moving stream, it may be an indication of pollution of some kind entering the water. Ponds and swamps usually have a higher ammonium nitrogen level than fast-flowing water. While levels of ammonium nitrogen in drinking water should not exceed 0.5 mg/L, streams or ponds near heavily fertilized fields may have higher concentrations of this ion.

Appendix 1-6

D. Chloride Ion, Cl^-

Chloride ion is one of the major negative ions in saltwater and freshwater. It originates from the dissociation of salts in water. There are many possible sources of manmade salts that may contribute to elevated chloride readings. Sodium chloride and calcium chloride, used to salt roads, contribute to elevated chloride levels in streams. Chlorinated drinking water and water softeners often increase chloride levels in wastewater of a community.

Sources of Chloride Ions
• River streambeds with salt-containing minerals
• Runoff from salted roads
• Irrigation water returned to streams
• Mixing of seawater with freshwater
• Chlorinated drinking water
• Water softener regeneration

Salinity, the total of all non-carbonate salts dissolved in water, is of interest in bodies of water where seawater mixes with freshwater, since aquatic organisms have varying abilities to survive and thrive at different salinity levels.

Expected Levels

The chloride level of water in freshwater streams and lakes can range from 0 to 250 mg/L. The recommended maximum level of chloride in drinking water is 250 mg/L.

E. pH

All water (even otherwise "pure" water) contains both hydrogen ions, H^+, and hydroxide ions, OH^-. The relative concentrations of these two ions determine the pH value.

On a pH scale of 0 to 14, a value of 0 is the most acidic, and 14 the most basic.	If pH = 7, the solution is NEUTRAL. If pH < 7, the solution is ACIDIC. If pH > 7, the solution is BASIC.

The pH scale is base-10 logarithmic, meaning that each unit change in pH corresponds to a tenfold change in acidity. (Another logarithmic scale you might be familiar with is the Richter scale used to measure earthquake intensity.) For example, a change from pH 7 to pH 8 represents a ten-fold decrease in acidity.

Factors that Affect pH Levels
• Acidic rainfall
• Algal blooms
• Level of hard-water minerals
• Releases from industrial processes
• Carbonic acid from respiration or decomposition
• Oxidation of sulfides in sediments

Rainfall generally has a pH value between 5.0 and 6.5 because of dissolved carbon dioxide and air pollutants. If the rainwater flows over soil containing hard-water minerals, its pH usually increases. As a result, streams and lakes are often basic, with pH values between 7.0 and 8.5.

Recreational fishermen care about pH because they know that trout and various kinds of nymphs can only survive in waters between pH 7 and 9.

Appendix 1-7

Effects of Various pH Levels on Aquatic Life	
pH	Effect
3.0–4.0	Unlikely that fish can survive for more than a few hours in this range, although some plants and invertebrates can be found at pH levels this low.
4.0–4.5	All fish and most frogs & insects absent.
4.5–5.0	Mayfly and many other insects absent. Most fish eggs will not hatch.
5.0–5.5	Bottom-dwelling bacteria begin to die. Plankton begin to disappear. Snails and clams absent. Metals (aluminum, lead) normally trapped in sediments are released into the acidified water in forms toxic to aquatic life.
6.0–6.5	Freshwater shrimp absent. Unlikely to be directly harmful to fish.
6.5–8.2	**Optimal for most organisms.**
8.2–9.0	Unlikely to be directly harmful to fish, but indirect effects occur at this level due to chemical changes in the water.
9.0–10.5	Likely to be harmful to perch if present for long periods.
10.5–11.0	Prolonged exposure is lethal to carp, perch.
11.0–11.5	Rapidly lethal to all species of fish.

In the Northeastern United States and Eastern Canada, fish populations in some lakes have been significantly lowered due to the acidity of the water caused by acidic rainfall. If the water is very acidic, heavy metals may be released into the water and can accumulate on the gills of fish or cause deformities that reduce the likelihood of survival. In some cases, older fish may live but be unable to reproduce because of the sensitivity of the reproductive portion of the growth cycle.

Expected Levels

The pH values of streams and lakes are usually between 7 and 8. In 2009, the pH of Akron's drinking water ranged between 6.99 and 7.96. "Harder" water (with high calcium ion concentration) often has pH values between 7.5 and 8.5.

F. Total Dissolved Solids (TDS)

Solids are found in streams and other natural water sources in two forms, *suspended* and *dissolved*. Suspended solids include silt, stirred-up bottom sediment, decaying plant matter, or sewage-treatment effluent. Suspended solids will not pass through a filter, whereas dissolved solids will.

TDS values will change when introduced to water from salts, acids, bases, hard-water minerals, or soluble gases that ionize in solution. The tests you perform will not tell you the *specific* ion responsible for the increase or decrease in TDS. They just give a general indication of the level of dissolved solids in a sample.

If TDS levels are high, especially due to dissolved salts, many forms of aquatic life are affected. High concentrations of dissolved solids can add a laxative effect to water or cause the water to have an unpleasant mineral taste. Dissolved ions may affect the acidity of a body of water, which in turn may influence the health of aquatic species. If high TDS readings are due to hard-water ions, then soaps may be less effective, or significant boiler plating may occur in heating pipes.

Sources of Total Solids
- Soil erosion
  - silt and clay
  - dissolved minerals
- Agricultural runoff
  - fertilizers
  - pesticides
- Urban runoff
  - road grime
  - rooftops
  - parking lots
  - road salt
- Industrial waste
  - dissolved salts
  - sewage treatment effluent
  - particulates
- Acid rain
- Organics
  - microorganisms
  - decaying plants and animals
  - gasoline or oil from roads

Appendix 1-8

Ions in Dissolved Solids and Their Sources	
Ions	**Source**
• calcium ion, Ca^{2+} • magnesium ion, Mg^{2+} • bicarbonate ion, HCO_3^-	• Hard water
• ammonium ion, NH_4^+ • nitrate ion, NO_3^- • phosphate ion, PO_4^{3-} • sulfate ion, SO_4^{2-}	• Fertilizer in agricultural runoff
• sodium ion, Na^+ • chloride ion, Cl^-	• Urban runoff
• sodium ion, Na^+ • potassium ion, K^+ • chloride ion, Cl^-	• Tidal mixing, minerals, or returned irrigation water
• hydrogen ion, H^+ • nitrate ion, NO_3^- • sulfite ion, SO_3^{2-} • sulfate ion, SO_4^{2-}	• Acidic rainfall

Expected Levels

TDS values in lakes and streams are typically found to be in the range of 50 to 250 mg/L. In areas of especially hard water or high salinity, TDS values may be as high as 500 mg/L. Drinking water will tend to be 25 to 500 mg/L TDS. United States Drinking Water Standards established by 1986 Amendments to the Safe Drinking Water Act include a recommendation that TDS in drinking water should not exceed 500 mg/L TDS. Fresh distilled water, by comparison, will usually have a value of 0.5 to 1.5 mg/L TDS.

Appendix 1-9

Drinking Water Samples

On Tuesday, October 12, you will pick up a sample bottle to collect a sample of drinking water. On Wednesday evening, you will collect a sample of drinking water to analyze in the chemistry lab. On Thursday, we will go to the chemistry lab to conduct further experiments and will then determine how it measures up as drinking water.

When you come to the chemistry lab on Thursday, your team will start the chemical analysis at one of six stations (A–F) to which you will be assigned. You will rotate through all of the remaining stations to complete the analysis of your water.

You will have 10 minutes at each station. You will follow these same steps at each station to analyze a sample for each of the six important water quality measures.

1. After gently shaking your container, pour a small amount of the water sample into a clean, dry beaker.
2. Place the tip of the sensor into the sample beaker.
3. <u>Hold the probe tip still</u> for at least 10 seconds. Record in the Data Table the reading displayed on the screen.
4. Repeat data collection on a second portion of your water sample, repeating steps 1-3. You should expect your two readings to be close to each other, but not identical. To minimize the effect of readings that vary too much, take a third and a fourth measurement.
5. Either now or as soon as possible after leaving the station as time permits, average your four readings and record the average along with the individual readings.

Repeat these steps for each sample that your group receives.

Appendix 1-10

Sample No. _____ Source _____

Data

Station		A	B	C	D	E	F
		Calcium Ion (Ca^{2+}) mg/L	Nitrate Ion (NO_3^-) mg/L	Ammonium Ion (NH_4^+) mg/L	Chloride Ion (Cl^-) mg/L	pH	Total Dissolved Solids (TDS) mg/L
Readings	1						
	2						
	3						
	4						
Average of four readings							

Comparison of Our Group's Findings with Expected Levels

	Calcium Ion (Ca^{2+}) mg/L	Nitrate Ion (NO_3^-) mg/L	Ammonium Ion (NH_4^+) mg/L	Chloride Ion (Cl^-) mg/L	pH	Total Dissolved Solids (TDS) mg/L
Expected level						
Our measured level						
Difference (+ or –)						
Notes						
Conclusions						

Appendix 1-11

Sample No._____ Source_____

Data

Station		A	B	C	D	E	F
		Calcium Ion (Ca^{2+}) mg/L	Nitrate Ion (NO_3^-) mg/L	Ammonium Ion (NH_4^+) mg/L	Chloride Ion (Cl^-) mg/L	pH	Total Dissolved Solids (TDS) mg/L
Readings	1						
	2						
	3						
	4						
Average of four readings							

Comparison of Our Group's Findings with Expected Levels

	Calcium Ion (Ca^{2+}) mg/L	Nitrate Ion (NO_3^-) mg/L	Ammonium Ion (NH_4^+) mg/L	Chloride Ion (Cl^-) mg/L	pH	Total Dissolved Solids (TDS) mg/L
Expected level						
Our measured level						
Difference (+ or −)						
Notes						
Conclusions						

Appendix 1-12

Appendix 2

The Mathematics of Water Usage

Over the last four days, you have maintained a Water Log where you recorded various ways in which you use water from various sources. In addition, you recorded the destination of this water it flowed through your body, over your car, across your dishes, into your lawn, etc. Now, we will work together to come to some conclusions.

To model the situation, we will do as many scientists do and make some assumptions that will simplify our work (and hopefully still be fairly accurate).

- Soda and coffee/tea products that you consumed (bottled or in a cup) were produced with water from Akron.
- Bottled water that you consumed came from another place besides Akron.

How much water do you use?

Step 1: In the following table, enter your totals for four days of water use.

Drinks			
total number of drinks _____ ×	Number of fluid ounces per drink _____	× $\frac{1}{128}$ gallon per ounce	= _____ gallons
Personal Hygiene			
total number of showers _____ ×	Number of minutes per shower _____	× 2.5 gallons per minute for average shower head	= _____ gallons
total number of baths _____		× 11 gallons per bath	= _____ gallons
total number of handwashings & toothbrushings _____ ×	Number of minutes running water per activity * _____	× 2 gallons per minute for average indoor faucet	= _____ gallons
Laundry			
total number of loads _____		× 43 gallons per load for average top-load washer	= _____ gallons
Toilet Use			
total number of flushes _____		× 1.6 gallons per flush for modern toilets	= _____ gallons
Lawn Sprinkler			
total number of waterings _____ ×	Number of minutes per watering _____	× 15 gallons per minute for average sprinkler	= _____ gallons
Lawn/Car via Hose			
total number of usages _____ ×	Number of minutes per usage _____	× 2 gallons per minute for average outdoor spigot	= _____ gallons
Washing Dishes by Hand			
total number of times _____ ×	Number of minutes running water per washing _____	× 2 gallons per minute for average indoor faucet	= _____ gallons
Dishwasher			
total number of loads _____		× 15 gallons per load for average dishwasher	= _____ gallons
Food Scrubbing			
total number of times _____ ×	Number of minutes running water per scrubbing* _____	× 2 gallons per minute for average indoor faucet	= _____ gallons
		total number of gallons of water used over 4-day period	_____ gallons

*If you recorded in seconds, multiply seconds by $\frac{1}{60}$ to convert to minutes.

September 16, 2010   GATE Learning Community

Step 2: What is the average number of gallons of water you used per day?

♣ your _average_ number of gallons of water used _per day_ [_____] gallons

Step 3: Looking at the LC students as a group, what was the average number of gallons each student used per day?

♦ class _average_ number of gallons of water used per student _per day_ [_____] gallons

♣ ♦ These averages will be used in a later comparison.

How much water is used by UA students?

Part 1. Dorm Residents

Now, you will try to estimate how much water the UA resident student body uses per day for drinks, personal hygiene, and toilet use. To start, those of you who live in UA dorms will calculate your daily water use for drinks, personal hygiene, and toilet use—see your numbers from Step 1 above. Add those amounts and divide this total by 4 to get your average daily use for these purposes.

your _average_ number of gallons of water used _per day_ for drinks, hygiene, and toilet use [_____] gallons

Assuming that your average daily use is similar to that of your fellow students, multiply your average by 3400 (the number of on-campus resident undergraduate students) to get an estimate of the amount of water used for these purposes per day by the students who live on campus.

estimated _total_ number of gallons of water used per day by UA resident students [_____] gallons

Part 2. Commuter Students

It is a more challenging problem to come up with a model to estimate how much water is used per day by commuting students. We will restrict ourselves to considering full-time undergraduate students who commute to campus rather than residing here. By some estimates, this population is about 16,000 students. These students drink water and coffee from campus vendors, use the bathrooms and wash their hands, but probably don't regularly brush their teeth or take showers on campus unless they are using the Rec Center. List some factors you would build into a model to try to determine the water usage of this group of students.

-
-
-

Using these factors, what is your estimate of the amount of water used per day for the above-listed purposes by commuter students? [_____] gallons

September 16, 2010   GATE Learning Community

According to Frank Horn, who spoke about UA green initiatives at last night's *Rethinking STEMM* program, in FY 2009 the University used 19,612,400 cubic feet of water that entered the sewer system, and 172,500 cubic feet of water for outdoor uses that did not enter the sewer system. The University paid approximately 5 cents per cubic foot for the water that was used and later entered the sewer system, and approximately 2.5 cents per cubic foot for the water that was used outdoors that did not enter the sewer system. What was UA's total water bill for FY 2009?

estimate of UA's total water bill for FY 2009	$

Let's put the price of a cubic foot of water in perspective. Do you think 5¢ per cubic foot of water is outrageously expensive? a bargain? dirt cheap? Let's see how it compares to the price of the following.

- 20-ounce bottle of water from UA vending machine ($1.50):

$$\frac{\$1.50}{20 \text{ oz.}} \times \frac{128 \text{ oz.}}{1 \text{ gal.}} \times \frac{7.48 \text{ gal.}}{1 \text{ ft}^3} \times \frac{100¢}{\$1} = 7200¢ \text{ per cubic foot (i.e., \$72/cubic foot!)}$$

- case of Giant Eagle bottled water (24 16.9-ounce bottles; $2.99 on sale): **710¢ per cubic foot**
- 32-ounce fountain Pepsi from Sizzling Zone ($1.80): **5400¢ per cubic foot**
- gallon of Acme lowfat milk ($1.99 on sale): **1500¢ per cubic foot**

How much water is used by a household?

For the next part of the project, we will look at household water usage. The class will divide into four groups. Each group will receive copies of three actual City of Akron water bills. On each water bill, **CONS 100 CF** shows the water consumption in hundreds of cubic feet. Each bill shows 7 CCF. How many cubic feet of water is that? (1 CCF = 100 ft^3)

We are more accustomed to measuring liquid volumes in liters, quarts, and gallons than in cubic feet. How many gallons of water are in 7 CCF? (1 CCF = 748 gal)

For each bill, what was the average number of gallons of water per day used <u>by each of the 3 residents</u> of the dwelling? Compare this number to <u>your</u> average daily water use (♣) and the average daily water use <u>per LC student</u> (♦).

For each bill, what rate (in $/CCF) is the homeowner paying for water and sewer? How have the rates changed?

estimate of homeowner's water & sewer rates	water rate		sewer rate	
Bill #1	$	per CCF	$	per CCF
Bill #2	$	per CCF	$	per CCF
Bill #3	$	per CCF	$	per CCF

Suppose you had your own home in Akron and used on average the amount of water per day that you calculated previously (♣). What would your total water and sewer bill be?

estimate of your total water bill	$

September 16, 2010   GATE Learning Community

Conclusion

This activity has focused on the economic aspect of the civic issue you are studying in the GATE Learning Community. As you prepare to discuss this aspect of the civic issue in your portfolio, it might help to think about what some economists call the water-diamond paradox.

The Water-Diamond Paradox

- Which is more important to your daily existence: water or diamonds?
- Which is more important to society's existence: water or diamonds?
- Which is more plentiful: water or diamonds?
- Which is less expensive: water or diamonds?
- If you could obtain either all of the water in the world or all of the diamonds in the world, which choice would you make in order to survive?
- If you were given the choice of receiving one additional gallon of water or one additional diamond, which would you rather have?
- Concerning water and diamonds: does price measure usefulness?

Nothing is more useful than water: but it will purchase scarce any thing; scarce any thing can be had in exchange for it. A diamond, on the contrary, has scarce any value in use; but a very great quantity of other goods may frequently be had in exchange for it.

–Adam Smith, in *The Wealth of Nations*, originally published in 1776.

Thanks to Dr. Donovan's microeconomics professor at the University of Delaware, who introduced him to the water-diamond paradox in McConnell, C.R., & Brue, S.L., *Microeconomics*, McGraw-Hill, 12th Ed., p. 131.

Appendix 3

18. Math and Economics of Water Usage

Keeping a Water Log

Over a <u>four-day time span</u> during the next week, including at least one weekend day, you will maintain a Water Log where you record various ways in which you use water from various sources. In addition, you will record the destination of this water as it flows through your body, over your car, across your dishes, into your lawn, etc. This sample page will offer guidance.

H₂O Usage				
	Number used	**Source**	**Time or volume used**	**Destination**
Drink	- 2 glasses - 1 sports-bottle - 1 glass	- Parent's / Wadsworth - on campus - Apt. / Akron	- 16oz. each - 32oz. - 12oz.	- Absorbed into body, eventually excreted - toilet/city sewer
Personal Hygiene				
Shower	- 1 shower	- Apt. / Akron	- 10 min.	- city sewer
Bath				
Handwashings & toothbrushings	- 7 handwashings - 3 toothbrushings	- Apt. / Akron - Apt. / Akron	- 30 seconds each - 5 seconds each	- city sewer
Other indoor water use				
Laundry	- 2 loads	- Laundromat / Akron		- city sewer
Toilet	- 2 flushes - 1 flushes	- Apt. / Akron - parent's / Wadsworth		- city sewer - parent's septic system
Washing dishes by hand	- 2 times	- Apt. / Akron	- 3 min. total each time	- city sewer
Dishwasher				
Food prep/scrubbing	- 1 time	- Apt. / Akron	- 30 seconds total	- city sewer
Outdoor water use				
Lawn sprinkler				
Lawn/car via hose	- washed car with parent's hose		- 20 min	- run-off into gravel driveway

99

52

Name_____ Water Log Day 1: Date_____

H₂O Usage				
	# used	Source	Time or volume used	Destination
Drink				
Personal Hygiene				
Shower				
Bath				
Handwashings & toothbrushings				
Other indoor water use				
Laundry				
Toilet				
Washing dishes by hand				
Dishwasher				
Food prep/scrubbing				
Outdoor water use				
Lawn sprinkler				
Lawn/car via hose				

Name_____   Water Log Day 2: Date_____

H₂O Usage				
	# used	Source	Time or volume used	Destination
Drink				
Personal Hygiene				
Shower				
Bath				
Handwashings & toothbrushings				
Other indoor water use				
Laundry				
Toilet				
Washing dishes by hand				
Dishwasher				
Food prep/scrubbing				
Outdoor water use				
Lawn sprinkler				
Lawn/car via hose				

Name_____   Water Log Day 3: Date_____

H$_2$O Usage				
	# used	Source	Time or volume used	Destination
Drink				
Personal Hygiene				
Shower				
Bath				
Handwashings & toothbrushings				
Other indoor water use				
Laundry				
Toilet				
Washing dishes by hand				
Dishwasher				
Food prep/scrubbing				
Outdoor water use				
Lawn sprinkler				
Lawn/car via hose				

Name_____ Water Log Day 4: Date_____

H₂O Usage				
	# used	Source	Time or volume used	Destination
Drink				
Personal Hygiene				
Shower				
Bath				
Handwashings & toothbrushings				
Other indoor water use				
Laundry				
Toilet				
Washing dishes by hand				
Dishwasher				
Food prep/scrubbing				
Outdoor water use				
Lawn sprinkler				
Lawn/car via hose				

The Mathematics of Water Usage

Over four days, you have maintained a Water Log where you recorded various ways in which you use water from various sources. In addition, you recorded the destination of this water it flowed through your body, over your car, across your dishes, into your lawn, etc. Now, we will work together to come to some conclusions.

To model the situation, we will do as many scientists do and make some assumptions that will simplify our work (and hopefully still be fairly accurate).

- Soda and coffee/tea products that you consumed (bottled or in a cup) were produced with water from Akron.
- Bottled water that you consumed came from another place besides Akron.

How much water do you use?

Step 1: In the following table, enter your totals for four days of water use.

Drinks			
total number of drinks _____	× Number of fluid ounces per drink _____	× $\frac{1}{128}$ gallon per ounce	= _____ gallons
Personal Hygiene			
total number of showers _____	× Number of minutes per shower _____	× 2.5 gallons per minute for average shower head	= _____ gallons
total number of baths _____		× 11 gallons per bath	= _____ gallons
total number of handwashings & toothbrushings _____	× Number of minutes running water per activity * _____	× 2 gallons per minute for average indoor faucet	= _____ gallons
Laundry			
total number of loads _____		× 43 gallons per load for average top-load washer	= _____ gallons
Toilet Use			
total number of flushes _____		× 1.6 gallons per flush for modern toilets	= _____ gallons
Lawn Sprinkler			
total number of waterings _____	× Number of minutes per watering _____	× 15 gallons per minute for average sprinkler	= _____ gallons
Lawn/Car via Hose			
total number of usages _____	× Number of minutes per usage _____	× 2 gallons per minute for average outdoor spigot	= _____ gallons

105

Washing Dishes by Hand			
total number of times _____	× Number of minutes running water per washing _____	× 2 gallons per minute for average indoor faucet	= _____ gallons
Dishwasher			
total number of loads _____		× 15 gallons per load for average dishwasher	= _____ gallons
Food Scrubbing			
total number of times _____	× Number of minutes running water per scrubbing* _____	× 2 gallons per minute for average indoor faucet	= _____ gallons
		total number of gallons of water used over 4-day period	_____ gallons

*If you recorded in seconds, multiply seconds by $\frac{1}{60}$ to convert to minutes.

Step 2: What is the average number of gallons of water you used per day?

♣ **your** average number of gallons of water used per day _____ gallons

Step 3: Looking at the class as a group, what was the average number of gallons each student used per day?

♦ **class** average number of gallons of water used per student per day _____ gallons

♣ ♦ These averages will be used in a later comparison.

106

How much water is used by UA students?

Part 1. Dorm Residents

Now, you will try to estimate how much water the UA resident student body uses per day for drinks, personal hygiene, and toilet use. To start, those of you who live in UA dorms will calculate your daily water use for drinks, personal hygiene, and toilet use—see your numbers from Step 1 above. Add those amounts and divide this total by 4 to get your average daily use for these purposes.

your average number of gallons of water used per day for drinks, hygiene, and toilet use	gallons

Assuming that your average daily use is similar to that of your fellow students, multiply your average by 3400 (the number of on-campus resident undergraduate students) to get an estimate of the amount of water used for these purposes per day by the students who live on campus.

estimated total number of gallons of water used per day by UA resident students	gallons

Part 2. Commuter Students

It is a more challenging problem to come up with a model to estimate how much water is used per day by commuting students. We will restrict ourselves to considering full-time undergraduate students who commute to campus rather than residing here. By some estimates, this population is about 16,000 students. These students drink water and coffee from campus vendors, use the bathrooms and wash their hands, but probably don't regularly brush their teeth or take showers on campus unless they are using the Rec Center. List some factors you would build into a model to try to determine the water usage of this group of students.

-
-
-

Using these factors, what is your estimate of the amount of water used per day for the above-listed purposes by commuter students?	gallons

In FY 2009 the University used 19,612,400 cubic feet of water that entered the sewer system, and 172,500 cubic feet of water for outdoor uses that did not enter the sewer system. The University paid approximately 5 cents per cubic foot for the water that was used and later entered the sewer system, and approximately 2.5 cents per cubic foot for the water that was used outdoors that did not enter the sewer system. What was UA's total water bill for FY 2009?

estimate of UA's total water bill for FY 2009	$

107

Let's put the price of a cubic foot of water in perspective. Do you think 5¢ per cubic foot of water is outrageously expensive? a bargain? dirt cheap? Let's see how it compares to the price of the following.

- 20-ounce bottle of water from UA vending machine ($1.50):
$$\frac{\$1.50}{20 \text{ oz.}} \times \frac{128 \text{ oz.}}{1 \text{ gal.}} \times \frac{7.48 \text{ gal.}}{1 \text{ ft}^3} \times \frac{100¢}{\$1} = 7200¢ \text{ per cubic foot (i.e., \$72/cubic foot!)}$$
- case of Giant Eagle bottled water (24 16.9-ounce bottles; $2.99 on sale): **710¢/cubic foot**
- 32-ounce fountain Coke from Union Market ($1.80): **5400¢/cubic foot**
- gallon of Acme lowfat milk ($2.29 on sale): **1700¢/cubic foot**

How much water is used by a household?

For the next part of the project, we will look at household water usage. The class will divide into groups. Each group will receive copies of five actual City of Akron water bills. On each water bill, **CONS. 100 CF** shows the water consumption in hundreds of cubic feet. Each bill shows 7 CCF. How many cubic feet of water is that? (1 CCF = 100 ft^3)

We are more accustomed to measuring liquid volumes in liters, quarts, and gallons than in cubic feet. How many gallons of water are in 7 CCF? (1 CCF = 748 gal)

For each bill, what was the average number of gallons of water per day used <u>by each of the 3 residents</u> of the dwelling?

Compare this number to <u>your</u> average daily water use (♣) and the average daily water use <u>per class student</u> (♦).

For each bill, what rate (in $/CCF) is the homeowner paying for water and sewer? How have the rates changed?

		estimate of homeowner's water & sewer rates		
		water rate		sewer rate
Bill #1	$	per CCF	$	per CCF
Bill #2	$	per CCF	$	per CCF
Bill #3	$	per CCF	$	per CCF
Bill #4	$	per CCF	$	per CCF
Bill #5	$	per CCF	$	per CCF

Suppose you had your own home in Akron and used on average the amount of water per day that you calculated previously (♣). What would your total water and sewer bill be?

estimate of your total water bill	$

Conclusion

The Water-Diamond Paradox

- Which is more important to your daily existence: water or diamonds?
- Which is more important to society's existence: water or diamonds?
- Which is more plentiful: water or diamonds?
- Which is less expensive: water or diamonds?
- If you could obtain either all of the water in the world or all of the diamonds in the world, which choice would you make in order to survive?
- If you were given the choice of receiving one additional gallon of water or one additional diamond, which would you rather have?
- Concerning water and diamonds: does price measure usefulness?

Nothing is more useful than water: but it will purchase scarce any thing; scarce any thing can be had in exchange for it. A diamond, on the contrary, has scarce any value in use; but a very great quantity of other goods may frequently be had in exchange for it.

--Adam Smith, in *The Wealth of Nations*, originally published in 1776.

Thanks to Dr. Donovan's microeconomics professor at the University of Delaware, who introduced him to the water-diamond paradox in McConnell, C.R., & Brue, S.L., *Microeconomics*, McGraw-Hill, 12th Ed., p. 131.

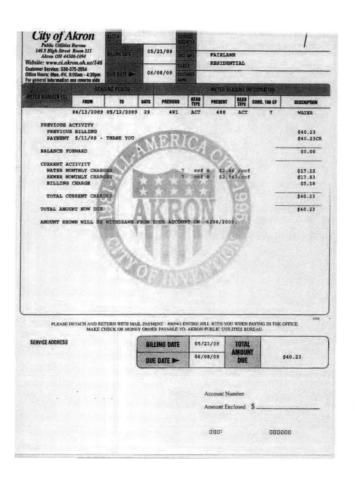

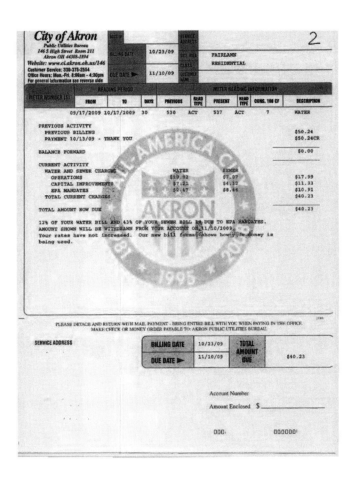

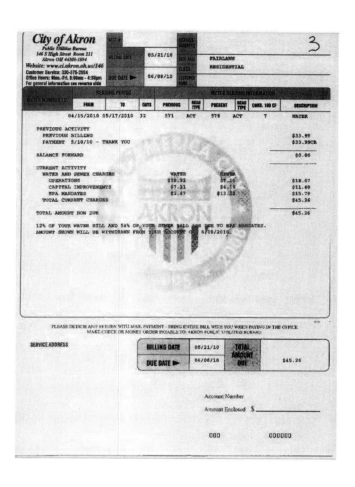

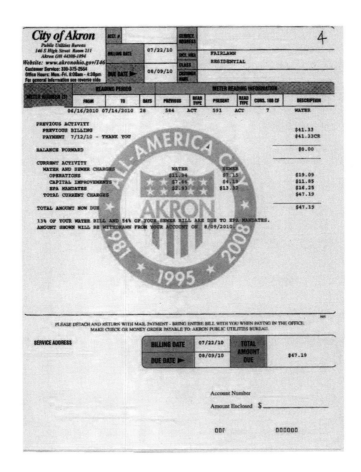

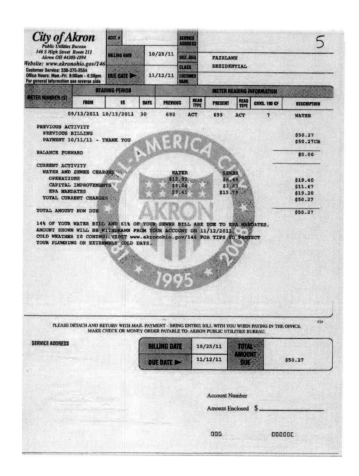

Chapter 4

Connections between Service Learning, Public Outreach, Environmental Awareness, and the Boy Scout Chemistry Merit Badge

Mark A. Benvenuto and Matthew J. Mio*

University of Detroit Mercy, Department of Chemistry and Biochemistry,
4001 W. McNichols Road, Detroit, Michigan 48221-3038
*E-mail: miomj@udmercy.edu.

> Helping Boy Scouts earn the chemistry merit badge can be used as a form of service learning for college and university students. Herein is described a short history of the merit badge, its multiple environmental connections, the specific requirements of the merit badge, some logistics for earning the badge and how to plan for a merit badge earning clinic. Past results are also explained, highlighting that the learning experience is beneficial for scouts and student volunteer/mentors alike.

Introduction

Many colleges and universities encourage some form of service as part of their students' formal educational requirements. In departments of chemistry, such service can take many forms. Most service opportunities fall under the broad category of service learning, in which college students, and faculty as well, engage in some service to their local community while being learners themselves. One course of action that involves service to a local community, environmental awareness, and basic chemistry is the chemistry merit badge of the Boy Scouts of America (BSA) (1).

The BSA merit badge program has an over 100-year history, with some badges surviving for almost the entire span of that time – such as Swimming or First Aid (1911) – and some badges only coming into existence in the relatively recent past, such as Composite Materials (2006) and Sustainability (2013). Several existing merit badges incorporate chemical science into all or part

© 2014 American Chemical Society

of their requirements, starting with the basic Chemistry badge. These others include: Composite Materials, Engineering, Energy, Environmental Science, Nuclear Science, Sustainability, and Pulp and Paper. Interestingly, there are several requirements within the Chemistry merit badge that have a distinctly environmental focus; that is, they are designed to raise environmental awareness on the part of the scout.

History

The chemistry merit badge is one of only 11 badges that have been in existence since the BSA program started in 1911 (*2*). Interestingly, the logo on the merit badge, a retort flask, has been preserved for the entire lifespan of the badge. The specific requirements of the badge, however, have changed slightly from one publication of the merit badge pamphlet to another. Merit badge pamphlets can be purchased as paper copies from all BSA Council Centers, but they can now also be downloaded as pdf documents (*3, 4*). For example, in the original set of merit badge requirements from 1911, scouts were asked for practical knowledge, as in "Why can baking soda be used to put out a small fire?" or "Give a test for chloride" or "What compound is formed when carbon is burned in air (*2*)?" As is evident above, many of the nine original merit badge requirements were directly related to the subjects of camping, hiking, agriculture and observation of nature. This is not surprising when the original purpose of the Boy Scouts (1910) was "to teach [boys] patriotism, courage, self-reliance, and kindred values (*5*)."

Environmental Connections and Modern Badge Requirements

In its modern form, the chemistry merit badge reflects the changing attitudes toward and applications of chemical science over the last 100 years. Most notably, the badge has several strong environmental connections, and therefore, touches upon the community-mindedness at the heart of service learning. As is customary, the requirements are stated in terms of duties/tasks that scouts must complete:

1.d. Discuss the safe storage of chemicals. How does the safe storage of chemicals apply to your home, your school, your community, and the environment?

6.a. Name two government agencies that are responsible for tracking the use of chemicals for commercial or industrial use. Pick one agency and briefly describe its responsibilities to the public and the environment.

6.b. Define pollution. Explain the chemical effects of ozone, global warming, and acid rain. Pick a current environmental problem as an example. Briefly describe what people are doing to resolve this hazard and to increase understanding of the problem.

6.c. Using reasons from chemistry, describe the effect on the environment of ONE of the following: 1) The production of aluminum cans or plastic milk cartons, 2) sulfur from burning coal, 3) used motor oil, 4) newspaper.

6.d. Briefly describe the purpose of phosphates in fertilizer and laundry detergent. Explain how the use of phosphates in fertilizers affects the environment. Also, explain why phosphates have been removed from laundry detergents.

7.d. Visit a county farm agency or similar governmental agency and learn how chemistry is used to meet the needs of agriculture in your county.

The other requirements for the badge focus on both safety and the five broad branches of the science, namely: analytical (measuring phenomena), biochemistry (life's basic mechanisms), inorganic (metallic and nonmetallic), organic (compounds of carbon), and physical (fundamental forces) chemistry.

Logistics of Earning the Badge

Merit badges are a modern form of a mentor and student learning in a one-on-one situation, but the BSA recognizes that some badges are more easily learned and earned in a professional setting than in a home or summer camp environment. Chemistry is one of those badges, and thus it is wise to use a laboratory as the setting for the session or clinic when the badge is to be earned. A visit to an academic or industrial lab is indeed one of the primary requirements of the modern chemistry merit badge. If one is planning to host a clinic at their place of work in off-hours, it's important to note that teaching labs often have more room than research labs, and thus work well when a large number of scouts attend the clinic. The actual number of scouts is best determined by the individuals running the clinic, and should always be a number that is safe, and with which the organizers/volunteers are comfortable.

Involving co-workers and representatives of your university's American Chemical Society (ACS) Student Members chapter is an excellent way to ensure that an appropriate number of people are involved in this sort of service learning and outreach activity. In turn, connecting directly with scout troops is often the quickest way to gauge interest in such a merit badge in your immediate area (6).

The requirements for the chemistry merit badge can be divided up into work stations or sites within a lab, so that all scouts are not working on the same requirement at the same time, and so that members of the service-learning team are not performing uneven amounts of work. Some of the requirements need water and other lab apparatus, while others simply require a spot for the scout and a member of the teaching team to sit and discuss an idea. Requirements 1 and 6, mentioned above, are two examples of the latter.

In the set-up phase, it is also important to determine how many mentors or instructors will be needed. Having a supply of merit badge pamphlets on hand helps to alleviate any concerns on the part of a college student volunteer. Often, mentors are afraid that they will not know an answer to a scout's question, and this provides a framework for the subject matter to be covered at each station.

The BSA require that at least one mentor be 21 years old or older, and be a certified merit badge counselor. The training required to be a certified counselor usually lasts one to two hours, and is conducted through a BSA council, or perhaps through the council's district training chairperson. Such training heavily emphasizes what is called the "No more, no less" idea. This means you cannot require a scout to do more than what is prescribed in the merit badge pamphlet, but that you also should allow no less than what the pamphlet states. This four-word slogan helps to keep merit badge requirements constant for every Boy Scout nationally.

Planning for a Chemistry Merit Badge Clinic

The following steps have proven remarkably useful to the authors in implementing chemistry merit badge clinics over the last 15 years:

- Connecting to local scout troops through student members of the ACS Student Members Chapter, or through faculty member contacts. Merit badge clinics, especially those in technical fields, are highly desired due to their perceived uncommonness. Some local science centers will even host clinics and charge for scouts to attend! Choose a suitable cap for your space and don't let high demand trump safety standards.
- Ensure you have at least one BSA certified counselor who will be present at the merit badge clinic. This is the person who signs and certifies that the scouts have performed the duties necessary to earn the merit badge.
- Inform the scout troops with whom the college or university has connected at least three months prior to the date of the clinic. This allows the scout troop to schedule the event on their training calendar and parents/guardians to plan ahead.
- Have a dress rehearsal in the lab one week prior to the event. Like its close twin, the chemical demonstration show, a merit badge clinic runs more smoothly with practice.
- Have enough safety goggles for each scout, each volunteer-mentor, and any parents who wish to be in the lab observing. Don't forget to have a number of small-sized goggles on hand for 10-11-year-old scouts.
- At the dress rehearsal, determine what equipment and supplies are needed at each station, and how many mentor-volunteers are needed at each station. Have enough extra material that you can accommodate spills and other unforeseen difficulties.
- Also at the dress rehearsal, have ACS Student Chapter members stake some sort of claim to specific stations or requirements, so that they can be sure they know the details of that specific task. Have students practice with peers as stand-in scouts.
- Ensure there is at least one extra mentor-volunteer who has no specific station, so that this person can continuously walk throughout the lab to ensure safety protocols are always met. Encourage student volunteers to treat their experience as that of a teaching assistant in a laboratory course.

- Designate a spot or room for a scout to read up on a requirement if it is obvious they have not done any study prior to attending the merit badge clinic. While not common, assuring scouts that second-chances are OK helps to alleviate a "test-taking" atmosphere.

While some of these may seem like trivial items for a to-do list, remember that the BSA is an organization for boys from ages 10 – 17, and that there are large learning differences between those ages. Spills do occur, and some scouts (not always the youngest ones!) attend without having had time to fully prepare. The idea is always to generate a positive learning experience for each scout.

To create a positive service learning experience for your student volunteer-mentors, the best place to start is their confidence. Young scouts can ask out-of-the-ordinary questions, so preparing for the unexpected is definitely in order. Discussion of the overall nature of the chemistry merit badge requirements can help ground a volunteer in the purpose of the clinic, which again has many environmental overtones. Most student volunteer-mentors independently recognize the unique opportunity that these clinics bring to review previously-learned course content and refresh lab safety and technique. In addition, a clinic wrap-up session can function as the reflection after this service-learning experience. Student volunteers often want to relate anecdotes about their interactions with the scouts, but this is only at the surface of the service they have provided. Asking volunteers to bring up the kinds of questions younger children ask about chemistry (versus study at the college level) and considering the implications of changing attitudes toward the utility of chemistry can facilitate rich conversation.

Conclusions

The BSA Chemistry merit badge can be an excellent form of service learning and public outreach for student groups at colleges and universities, one that has numerous and obvious environmental connections. An understanding of the badge requirements before starting is essential, but easy to obtain. When setting up what can be called a chemistry merit badge clinic, having at least one certified merit badge councilor is required. Planning ahead of time for the necessary space, materials, and equipment is important. Contacting troops approximately 3 – 6 months ahead of the event results in the most prepared scouts. Having enough personnel on hand to attend to each station while the merit badge clinic is running will ensure that the event goes safely and smoothly. And perhaps most importantly of all, ensure that both your own personnel and the scouts have fun while earning the badge! It can be a very rewarding experience for everyone involved.

References

1. Boy Scouts of America website, scouting.org (accessed 8 July 2014).
2. *Boy Scouts Handbook*, 1st ed.; 1911; gutenberg.org/files/29558/29558-h/29558-h.htm (accessed 8 July 2014).

3. Chemistry merit badge pamphlet website, rpadden.com/BSA/m-b/0018.pdf (accessed 8 July 2014).
4. Chemistry merit badge requirements website, meritbadge.org/wiki/images/0/0c/Chemistry.pdf (accessed 8 July 2014)
5. *Boy Scouts of America Wikipedia*; en.wikipedia.org/wiki/Boy_Scouts_of_America (accessed 8 July 2014).
6. Melzer, M.; Wiley, K. Fun, Fire and Food! How We Help Local Boy Scouts Earn Their Chemistry Merit Badges. *inChemistry* **2014**, *22*, 26–28.

Chapter 5

A Service-Learning Project Focused on the Theme of National Chemistry Week: "Energy-Now and Forever" for Students in a General, Organic, and Biological Chemistry Course

Elizabeth S. Roberts-Kirchhoff[*]

Department of Chemistry and Biochemistry, University of Detroit Mercy, 4001 W. McNichols Roadd, Detroit, Michigan 48221
*E-mail: robkires@udmercy.edu.

Each fall, the Detroit Section of the American Chemical Society (ACS) and the Girl Scouts of Southeastern Michigan sponsor a Chemistry Day for Junior Girl Scouts and their leaders. The goal of the event is to provide an opportunity for the participants to learn about the role of chemistry in everyday lives. The theme of the day is based on the theme of National Chemistry Week sponsored by the ACS. One of the learning outcomes for the students in a general, organic and biological chemistry course at the University of Detroit Mercy (UDM) is to discuss how concepts in the course relate to medicine, industry or environmental issues. The UDM students complete a service-learning project as one way to fulfill this outcome. The students, working in groups, designed posters which were displayed at the annual Chemistry Day based on the National Chemistry Week theme. For 2013, the theme was "Energy-Now and Forever" and each poster described one type of energy, basic chemical concepts involved, new and emerging technologies related to that energy source, and environmental impacts. The assignment instructions, evaluation rubric, and peer evaluation are discussed.

© 2014 American Chemical Society

Introduction

Service learning as defined by the American Chemical Society (ACS) is "an educational program that integrates community service with classroom instruction to enhance learning, teach civic responsibility and strengthen communities" (*1*). There is a significant body of literature indicating that service learning benefits faculty, students, institutions, and community partners (*2–6*). In the course of organizing and implementing the service-learning experiences, faculty benefit from working with more engaged students, and educational institutions benefit from strengthening their connections with community partners (*3*). For university students, service learning has been described by Kuh as one of the high-impact educational practices that leads to deeper learning and enhanced outcomes (*6*). Other high-impact educational practices for students include participation in learning communities, study abroad, and undergraduate research. These high-impact educational practices and their benefits for students were reviewed as part of the Liberal Education and American's Promise (LEAP) initiative (*6*) from the Association of American Colleges and Universities (AAC&U). The LEAP initiative is a ten-year national activity by the AAC&U to align goals for college learning with the changing global society. The student benefits of these high-impact educational practices include higher grade point averages and increased retention (*6*). Results show that involvement in these activities are beneficial for all students but historically underserved students benefit even more from these experiences as compared to majority students (*6*).

There are many examples of service learning in chemistry in the literature. These include service learning in general chemistry (*4, 7–12*), analytical chemistry (*13, 14*) and biochemistry (*15–17*) courses. Some of these examples are intense semester-long projects and others are simpler and shorter experiences. Some of these have the focus or are related to environmental chemistry (*8, 12–14*). In this example, the idea of the simple, short project idea as described by Sutheimer as a way to simplify service-learning efforts in chemistry (*12*) was used in an allied health chemistry course with the local section of the Girls Scouts as the community partner.

Course and Project Description

Chemistry 1015, Chemical Principles, is the first in a two-course sequence of general chemistry, organic chemistry and biochemistry for our allied health majors including nursing and dental hygiene majors. Chemistry 1015 is an introduction to chemistry including the topics of measurements, atomic and molecular structure, energy, gases, solutions, chemical quantities and reactions. The description of the course from the syllabus is "Chemistry 1015 is the first in a two-course series designed to introduce non-science majors and students in the health sciences and related fields to basic chemical principles. It is intended that this course, along with Chemistry 1025, will provide students with an understanding of fundamental chemical principles, an appreciation for the application of chemistry in modern society, and the skills to solve common problems in chemistry." The second course in the series is Chemistry 1025, Introductory Chemistry for the

Health Sciences. The topics covered in Chemistry 1025 include acid and base chemistry, nuclear chemistry, an introduction to the different classes of organic compounds applicable to biochemistry and their reactions, the structure and function of biological molecules, and the metabolic pathways involved in energy production. Both of these courses also fulfill a science core requirement for undergraduate students at the University of Detroit Mercy (UDM). Before this change, all students were required to take the two-part series, Chemistry 1010, Principles of Inorganic Chemistry, and Chemistry 1020, Principles of Organic and Biochemistry. After discussions with the faculty in the McCauley School of Nursing at UDM, the sequence was redesigned and the students were required to take a chemistry placement exam. Some of the topics that had been in the first course, acid and base chemistry and nuclear chemistry, were moved into the second course. The students take the ACS Toledo Placement Exam during their summer orientation. Depending on their score on this exam and their score on the math placement exam, they are placed into (1) a math class prior to college algebra and no chemistry course, (2) Chemistry 1015, or (3) Chemistry 1025. If students have had a least one semester of a college chemistry course or have advanced placement credit for chemistry, they also place into Chemistry 1025. In fall 2013, 85% of the students in the Chemistry 1015/1025 sequence were enrolled in Chemistry 1015. Thus, a majority of the students take the two-semester series.

One of the student learning outcomes in Chemistry 1015 is to discuss how the concepts in the course relate to medicine, industry or environmental issues. The UDM students complete a service-learning project as one way to fulfill this outcome. For several years, the students in CHM 1010 worked in groups and designed posters which were displayed at the annual Girl Scout Chemistry Day based on the ACS National Chemistry Week theme. In fall 2013, the theme was "Energy-Now and Forever" and the goal was to explore "the differences between renewable and nonrenewable energy sources and how each type of energy source powers our planet" (*18*). The students worked in groups of three or four and the poster content for each group was one energy source. There were nine total energy sources and these were the sources identified in the ACS National Chemistry Week materials: solar, water, geothermal, oil, nuclear, gas, wind, biomass, and batteries (*18*). The students were required to include the following information on their posters: describe what energy is, describe how it is produced, discuss emerging energy technologies related to this energy source, identify elements from the periodic table important for this energy source, and discuss environmental impacts. Each student completed a peer evaluation and a self-evaluation after completing the poster. The peer and self-evaluation was adapted from the *Peer Evaluation* from David Murray, Ph.D. from University of Buffalo School of Management (*19*). All students were encouraged to also complete a Civic Engagement Student Pre-Service Survey and Student Post-Service Survey modified from the Curriculum Tools from the American Association of Community Colleges (*20*).

During the second week of the course, students were given the description of the assignment. They were randomly assigned to groups of three or four and then assigned one of the nine energy sources. One month before the due date for the poster, each group submitted at least five references for approval by the instructor. Two weeks before the due date, the students submitted a rough draft. This needed to include the content (with references) and a rough layout of the material. Within several days, the instructor provided feedback and the trifold poster board. The students used the feedback to revise their poster prior to putting together the final project. The poster was submitted on a Friday, the day before the Girl Scout event. Each group presented their poster to the rest of the class during the recitation session on the Friday morning before the Chemistry Day. The peer evaluations were also due at this time.

Annual Chemistry Day

In November 2013, the 9th annual Girl Scout Chemistry Day was held in a suburb of Detroit, MI at the First United Methodist Church of Troy. The annual Girl Scout Chemistry Day is collaboration between the Detroit Section of the American Chemical Society and the Girl Scouts of Southeastern Michigan. In 2013, approximately 240 Junior Girl Scouts and their chaperones spent the day learning about the role of chemistry in everyday lives. There were about 50 volunteers including chemists from the Detroit Section of the ACS, students from the Detroit ACS student chapters from Wayne State University, UDM, and Oakland University, and other University students from UDM. The other students from UDM include some of the students in Chemistry 1015, the class that designed the posters for their service-learning project.

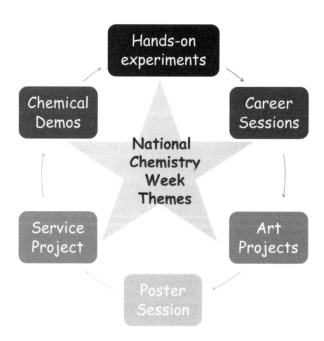

Figure 1. Activities for the annual Chemistry Day for Junior Girl Scouts.

There are several activities at the Chemistry Day and these are outlined in Figure 1. The events include registration, welcoming remarks and a continental breakfast, rotation through three major activities, closing remarks with a pizza lunch and liquid nitrogen ice cream. For the three major activities, the girl scouts rotate between hands-on-experiments, the career session, and an art project. These activities focus on the National Chemistry Week theme of energy.

The posters from the service-learning project prepared by the UDM students in Chemistry 1015 were displayed in the vestibule outside of the career session room and the art project room. This area is where the girl scouts and their chaperones checked in for registration and where they rotate through every event. The service project has often been the collection of nonperishable food items for a local food pantry. Each girl is encouraged to bring these with them and they are collected at the registration table. The chemical demonstrations are presented during the career session and during the beginning and ending of the day.

The theme of this Chemistry Day and the poster session correlates with one of the Junior Girl Scouts National Leadership Journey series. The National Program Portfolio for the Girl Scouts has two main parts-the National Leadership Journeys and *The Girl Scouts Guide to Scouting*. In the latter is found information on badges, girl scout history, tradition, and the girl scout handbook. The Leadership Journeys are aimed at developing and experiencing leadership through activities where the girls discover themselves, connect with others, and take action to make the world a better place. There are three series: It's Your world—Change It!, It's Your Planet—Love It!, and It's Your Story—Tell It! (*21*). These help junior girl scouts (4th -5th graders) "understand what it means to be a leader who makes a difference in the world through unique leadership and advocacy challenges" (*21*). In the series, It's Your Planet—Love It!, they earn three new leadership awards as they explore their own energy, the energy in their places and spaces (buildings), and the energy of getting from here to there (transportation) (*22*). The three awards are the Energize Award, the Investigate Award, and the Innovate Award. Some of the topics covered in this series include definition of energy, discussing the basic types of energy, energy use in buildings, energy audits, and conserving energy among others (*22*). By participating in the Chemistry day, the girls fulfilled several requirements towards their awards in the Journey series, It's Your Planet-Love It!

Results

On the day of the event, about 25% of the students in Chemistry 1015 attended as volunteers. The posters were placed in the vestibule where the registration occurs and outside the doors to the career session and art project. They were specifically used during the career session. In the career session, several women chemists talked about the role of a chemist in various aspects of the creation of a solar panel (polymer chemist, patent attorney, and teacher). Since the girl scouts are sitting during much of the 40-minute career session, they are given a break to get up and move outside the room to view the posters. The girl scouts viewed the posters and completed a matching activity (on paper) focused on chemical elements and energy sources. They put their names on these and entered them in a drawing for girl-scout or chemistry-themed prizes. The winning entries were drawn at random during lunch and the closing activities.

The Chemistry 1015 student scores for the posters were determined using the grading rubric shown in Table 1. The work before the final submission included the submission of references (1 point) and the rough draft (5 points). The poster was evaluated on the graphics (text, clarity, and originality), attractiveness, mechanics/grammar, poster content, and references. There were four levels of performance (exceptional, admirable, adequate and unacceptable) for each of the categories. There was also 1 point associated with the completion of the peer and self-evaluation. The results from the student scores on the poster submissions are shown in Figure 2. The groups were successful in creating attractive posters with both relevant and original graphics. The groups were also successful in preparing rough drafts of their posters and the list of five references submitted with the draft of the poster. The groups lost points on the references when the posters were evaluated (often not giving attribution throughout or only giving a website address as a citation), the poster message/content, grammar, and clarity of the text (text too small). The peer/self-evaluation (*19*) was given to the students with the assignment description and was reviewed by the instructor. Some students did not complete the self-evaluation portion of the evaluation. The score would only affect their own score and not the score for the entire group.

Each Chemistry 1015 student evaluated their partners and themselves on the quality of work, timeliness of work, task support, interaction, attendance, responsibility, involvement, leadership and the overall performance. These were scored on a five-point scale. Each student also listed the specific tasks they completed for the project. There was also a place where the students could write general comments about the project, the experience or working with their group. In this section, students could reflect on the project. Overall the students did very well on this project. The scores have improved over the past several years most likely due to the implementation three years ago of the requirement for the rough draft and the submission of the five references before the due date of the completed project. In the General Comments section, 19 of 33 students commented in some way that their group worked well together. Seven students commented on the fact that it was difficult to schedule meetings because of class schedules or students who were commuters. Surprisingly, only two students mentioned someone in their group not contributing in a substantial way to the project. Some representative comments from this section include:

> "We all worked and communicated efficiently and I feel we did an overall great job on our oil energy poster."
> "I thought our group worked really well together. I think one of our biggest challenges my group faced was scheduling time."
> "I really learned a lot about wind energy-it's a lot more than a windmill."
> "Great learning experience"

"It's hard to coordinate with 4 different people especially when 3 of us commute. Overall I thought our poster was good.
"Overall a great team effort."
"It was a nice experience and I made new friends."
"I thought this was an interesting project that allowed us to learn about energy in an easy and fun way. I also like that we are using the project to teach young girls about it too."
"It was very enlightening to learn so much about solar energy. I really enjoyed getting to know some of the fellow students in my class."

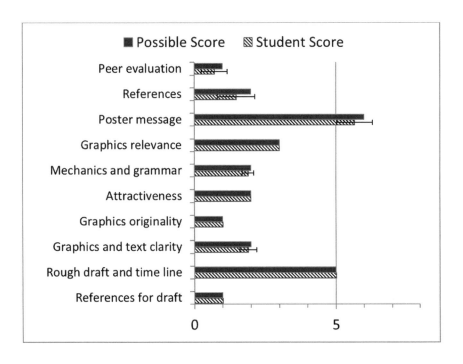

Figure 2. Results from the scoring of student posters. The solid black bars indicate the possible score while the hatched bar show the average + standard deviation for the group scores.

Table 1. Grading Rubric for Energy-Themed Posters

CATEGORY	Exceptional	Admirable	Adequate	Unacceptable
References (1 pt)	Five acceptable references are provided by the due date (1 pt)	Some of the references presented are acceptable (0.5 pt)	No acceptable references (0 pt)	No acceptable references (0 pt)
Rough Draft and Timeline (5 pts)	Draft contains all pertinent chemical content with references. Draft and poster turned in by due date. (5 pts)	Draft contains chemical content with references. Poster and draft turned in by due date. (3-4 pts)	Draft missing information or references. Poster or draft not turned in by due date. (1-2 pts)	Draft and poster not turned in by required date. (0 pts)
Graphics/Text-Clarity (2 pts)	Graphics are in focus and the content easily viewed and identified from 6 ft. away. (2 pts)	Most graphics are in focus and the content easily viewed and identified from 6 ft. away. (1 pt)	Most graphics are in focus and the content is easily viewed and identified from 4 ft. away. (1 pt)	Many graphics are not clear or are too small. (0 pts)
Graphics-Originality (1 pt)	Several of the graphics used on the poster reflect an exceptional degree of student creativity in their creation and/or display. (1 pt)	One or two of the graphics used on the poster reflect student creativity in their creation and/or display. (0.5 pt)	The graphics are made by the student, but are based on the designs or ideas of others. (0 pt)	No graphics made by the student are included. (0 pts)
Attractiveness (2 pts)	The poster is exceptionally attractive in terms of design, layout, and neatness. (2 pts)	The poster is attractive in terms of design, layout and neatness. (1 pts)	The poster is acceptably attractive though it may be a bit messy. (0.5 pt)	The poster is distractingly messy or very poorly designed. It is not attractive. (0 pts)
Mechanics and Grammar (2 pts)	Capitalization and punctuation are correct throughout the poster. There are no grammatical mistakes on the poster. (2 pts)	There is 1 error in capitalization or punctuation. There is 1 grammatical mistake on the poster. (1.5 pts)	There are 2 errors in capitalization or punctuation. There are 2 grammatical mistakes on the poster. (1 pt)	There are more than 2 errors in capitalization or punctuation. There are more than 2 grammatical mistakes on the poster. (0 pts)

CATEGORY	Exceptional	Admirable	Adequate	Unacceptable
Graphics-Relevance (3 pts)	All graphics are related to the topic and make it easier to understand. Graphics enhance the text. All borrowed graphics have a source citation. (3 pts)	All graphics are related to the topic and most make it easier to understand. All borrowed graphics have a source citation. (2 pts)	All graphics relate to the topic. Most borrowed graphics have a source citation. (1 pt)	Graphics do not relate to the topic OR several borrowed graphics do not have a source citation. (0 pts)
Poster Message/Content (6 pts)	Poster is fun, motivational, and highlights the chemistry and environmental impact of the topic. There is an emphasis on the science in the information given. The poster is appropriate for a junior girl scout (ages 8-11). (6 pts)	Poster highlights the chemistry with an emphasis on the science. The poster is appropriate for a junior girl scout (ages 8-11). (5-4 pts)	Poster highlights the chemistry. No clear emphasis on the science or environmental impact in the information. It is questionable if the poster is appropriate for a junior girl scout (ages 8-11). (3-2 pts)	No clear emphasis on the science in the information. It is questionable if the poster is appropriate for a junior girl scout (ages 8-11). (1 pt)
References on poster (2 pts)	Correctly cites all internet and print references. Attribution given throughout poster. Accurate and reliable sources used. (2 pts)	Cites all internet and print references. Attribution given throughout poster. Accurate and reliable sources used. (1 pts)	Cites references. Attribution missing for some information. Some unreliable references (0.5 pt)	Some information cited. Errors in format. Unreliable references (0 pts)
Peer Evaluation (1 pt)	Peer Evaluation is thoughtfully completed and turned in by due date. Information for all group members is included. (1 pt)	Peer Evaluation is not complete but is turned in by due date. Some information about group members is missing. (0 pts)	Peer evaluation not turned in on due date (0 pts)	No Peer evaluation done (0 pts)

Table 2. Results from the Pre-Service Survey and Post-Service Survey Modified from (20)

Survey question	Strongly Agree + Agree (%)		Disagree + Strongly Disagree (%)	
	Before	After	Before	After
I have a good understanding of the needs and problems facing the community in which I live.	80	94		
If everyone works together, many of society's problems can be solved.	100	90		
I have a responsibility to serve my community.	96	97		
I learn course content best when connections to real-life situations are made.	72	93		
The idea of combining course work with service to the community should be practiced in more courses at this college.	68	86		
I probably wont volunteer or participate in the community after this course ends.			72	77
The energy poster assignment helped me to understand some of the course material better.*		76		24
The energy poster assignment helped me to see how the subject matter I learned can be used in everyday life.*		83		17
The service I did by creating this energy poster was not at all beneficial to the community.*		17		83

* These questions were on the Post-Service Survey and not included on the Pre-Service Survey.

The students were also encouraged to complete a Civic Engagement Student Pre-Service Survey and Student Post-Service Survey modified from the Curriculum Tools from the American Association of Community Colleges (20). These were modified and converted to online surveys. The students were sent the link prior to the poster session and then sent the link after the poster session. For the post-service survey, I added three additional questions to the survey. The questions and the response rates are shown in Table 2. Some of the largest gains were seen when students were asked to respond to the statement, "I learn course content best when connections to real-life situations are made" with a 21% increase. The students liked the idea of combining course work with service to the community with an 18% increase. There was a decrease from 100 to 90 % in response to the statement: "If everyone works together, many of society's problems can be solved." Even though this was a decrease, a significant majority-90%- still think that working together will help in solving societal problems. Seventy-six percent of the students agreed or strongly agreed that the poster project helped them understand the course material better. Eighty-three percent of the students responding said they strongly agreed or agreed that they could see the application of the course material to their life.

Discussion and Summary

In a chemistry course for non-majors, UDM students completed a simple, short service-learning project to fulfill the student learning outcome: discuss how concepts in the course relate to medicine, industry or environmental issues. The students, working in groups, designed posters based on the National Chemistry Week theme which were displayed at the annual Chemistry Day. For 2013, the theme was "Energy-Now and Forever" and each poster described one type of energy, basic chemical concepts involved, new and emerging technologies related to that energy source, and environmental impacts.

Both direct and indirect measures were used to determine if the students met the learning outcome. The direct measure was the generation of the posters and these were evaluated using a scoring rubric. The students were able to gather information on energy, relevant to topics in general chemistry and the environment, and then present the information at a level appropriate for a junior girl scout. Weaknesses for some of the posters included not giving attribution for information throughout the poster, the readability of some of the text, and missing some required information. Results from the general comments section of the peer evaluation and the pre- and post-service surveys were used as indirect measures that students had met the learning outcome. Eighty-three percent of the students reported that the assignment helped them to see how the subject matter can be used in everyday life. The students in the course were enthusiastic about the project, and approximately seventy-five percent of the students reported that this project helped them understand some of the course material better. The students reported they learn course material better when connections to real life situations are made. In addition, the students worked well with their teams and even commented on the fact that they met some new friends. For these students, many who are commuters and in the first months of their college careers, they report not only a benefit of understanding the material but also of connecting with others in the class. The service-learning project was a benefit to the community partner since the posters provided content aligned with the Girl Scout Journey series, "It's Your Planet-Love it!". The girl scouts, through their exposure to the posters and the other events of the day, completed some of the requirements for the awards in the Journey series.

Acknowledgments

I thank the organizers of the Annual Chemistry Day for Girls Scouts especially Mary Kay Heidtke of the Detroit Section of the ACS and Caroline Feathers of the Girl Scouts of Southeastern Michigan.

References

1. *Service learning Resources for Chemistry Faculty* (n.d.); American Chemical Society http://www.acs.org/content/acs/en/education/resources/undergraduate/service-learning-resources-for-chemistry-faculty.html (accessed July 13, 2014).
2. Cavinato, A. G. Service Learning in Analytical Chemistry: Extending the Laboratory into the Community. In *Active Learning: Models from the Analytical Sciences*; Mabrouk, P. A., Ed.; ACS Symposium Series 970; American Chemical Society: Washington, DC, 2007; pp 109−122.
3. Zlotkowski, E. Introduction. In *Service-Learning and the First Year Experience: Preparing Students for Personal Success and Civic Responsibility* (Monograph 34); Zlotkowski, E., Ed.; University of South Carolina, National Resource Center for the First-Year Experience and Students in Transition: Columbia, SC, 2002; p x.
4. Esson, J. M.; Stevens-Truss, R.; Thomas, A. Service learning in Introductory Chemistry: Supplementing Chemistry Curriculum in Elementary Schools. *J. Chem. Educ.* **2005**, *82*, 1168–1173.
5. Vogelgesang, L. J.; Ikeda, E. K.; Gilmartin, S. K.; Keup, J. R. Service-Learning and the First-Year Experience: Outcomes Related to Learning and Persistence. In *Service-Learning and the First-Year Experience: Preparing Students for Personal Success and Civic Responsibility.* (Monograph 34); Zlotkowski, E., Ed.; University of South Carolina, National Resource Center for the First-Year Experience and Students in Transition: Columbia, SC, 2002; pp 15–24.
6. Kuh, G. D. *High-Impact Educational Practices: What They Are, Who Has Access to Them, and Why They Matter*; Association of American Colleges and Universities: Washington, DC, 2008.
7. Kalivas, J. H. A Service-Learning Project Based on a Research Supportive Format in the General Chemistry Laboratory. *J. Chem. Educ.* **2008**, *85*, 1410–1415.
8. Kesner, L; Eyring, E. Service Learning General Chemistry: Lead Paint Analyses. *J. Chem. Educ.* **1999**, *76*, 920–923.
9. Hatcher-Skeers, M.; Aragon, E. Combining Active Learning with Service Learning: A Student-Driven Demonstration Project. *J. Chem. Educ.* **2002**, *79*, 462–464.
10. Morgan Theall, R. A.; Bond, M. R. Incorporating Professional Service as a Component of General Chemistry Laboratory by Demonstrating Chemistry to Elementary Students. *J. Chem. Educ.* **2013**, *90*, 332–337.
11. Kuntzleman, T. S.; Baldwin, B. W. Adventures in Coaching Young Chemists. *J. Chem. Educ.* **2011**, *88*, 863–867.

12. Sutheimer, S. Strategies to Simplify Service-Learning Efforts in Chemistry. *J. Chem. Educ.* **2008**, *85*, 231–233.
13. Gardella, J. A.; Milillo, T. M.; Gaurav, O.; Manns, D. C.; Coeffey, E. Linking Advanced Public Service Learning and Community Participation with Environmental Analytical Chemistry: Lessons from Case Studies in Western New York. *Anal. Chem.* **2007**, *79*, 811–818.
14. Fitch, A.; Wang, Y.; Mellican, S.; Macha, S. Lead Lab Teaching Instrumentation with One Analyte. *Anal. Chem.* **1996**, *68*, 727A–731A.
15. Montgomery, B. L. Teaching the principles of biotechnology transfer: A Service-Learning Approach. *Electron. J. Biotechnol* **2003**, *6*, 13–15.
16. Grover, N. Introductory Course Based on a Single Problem: Learning Nucleic Acid and Biochemistry from AIDS Research. *Biochem. Mol. Biol. Educ.* **2004**, *32*, 367–372.
17. Harrison, M. A.; Dunbar, D.; Lopatto, D. Using Pamphlets to Teach Biochemistry: A Service-Learning Project. *J. Chem. Educ.* **2013**, *90*, 210–214.
18. *National Chemistry Week 2013* (n.d.); American Chemical Society http://www.acs.org/content/acs/en/education/outreach/ncw/past/ncw-2013.html (accessed May 27, 2014)
19. Murray, D. *MGS351 Course Website* (n.d.); University of Buffalo, mgt.buffalo.edu/departments/mss/djmurray/mgs351/PeerEval.doc (accessed May 2, 2012).
20. *Service Learning: Curriculum Tools* (n.d.); American Association of Community Colleges http://www.aacc.nche.edu/Resources/aaccprograms/horizons/Pages/curriculumtools.aspx (accessed March 2014)
21. *Welcome to the Journeys* (n.d.); Girls Scouts https://www.girlscouts.org/program/journeys/ (accessed June 20, 2014).
22. *It's Your Planet-Love It!: Introducing the Second Series of Journeys* (n.d.); Girl Scouts https://www.girlscouts.org/program/journeys/your_planet/ (accessed June 20, 2014).

Chapter 6

Service Learning in Environmental Chemistry: The Development of a Model Using Atmospheric Gas Concentrations and Energy Balance To Predict Global Temperatures with Detroit High School Students

Daniel B. Lawson*

University of Michigan-Dearborn, 4901 Evergreen Road,
University of Michigan-Dearborn, Dearborn, Michigan 48128-1491
*E-mail: dblawson@umich.edu.

This chapter entails the development of a fairly simplistic but remarkably accurate zero-dimensional energy balance model of the average global temperature with corrections for infrared (IR) active atmospheric gases developed when teaching the topic of "Scientific Modeling" to a group of high school teachers and students from several Detroit Public High Schools. The basis for teaching scientific modeling was part of a broader effort to bring opportunities in information technology within the context of science, technology, engineering and mathematics to a group of underrepresented and underserved urban high school students. The group approach to further developing the model was continued for an audience of graduate students in environmental studies taking a course in environmental chemistry. The resulting model accounts for thermal contributions of IR active gases in the atmosphere and predict the global average temperature to within -2 °C of the measured average. The model is also applied to Venus and Mars yielding average planetary temperatures that differ by 2.6% and -3.1%, respectively of the observed average temperatures. This model is unique in that it is constructed from molecular

© 2014 American Chemical Society

principles mostly taught at the general chemistry level and when applied in a group learning environment proves fairly straightforward to learn and present.

Introduction

This chapter describes the development of a fairly simplistic but remarkably accurate model of average global temperature developed when teaching the topic of "Scientific Modeling" to a group of high school teachers and students from several Detroit Public High Schools as part of a grant from the National Science Foundation (NSF) on learning using information technology (IT) within the context of science, technology, engineering and mathematics (STEM). This IT/STEM program utilized high school teachers along with university faculty, graduate and undergraduate students to guide high school students through the various STEM areas. The project was originally designed to increase the opportunities for underrepresented and underserved urban high-school students to learn, experience and use information technologies within the context of STEM. The project also provided opportunities for K-12 STEM teachers and post-secondary STEM content experts to learn about, design and deliver IT/STEM enrichment experiences and opportunities for students as well as develop a series of learning resources and deliverables that models the uses of IT within the context of STEM.

The project concentrated on all 4 areas of STEM creating 4 project-based design teams to address IT in science, engineering, technology and mathematics. Each team included 4 to 6 high school students, 1 K-12 teacher, 1 undergraduate/ graduate university student and 1 content area faculty expert. The *science* team concentrated on 3 different but related applications of IT in the sciences including measurement, modeling and mapping while providing experience using GIS, GPS for mapping and mathematical descriptions of locations; using temperature and light sensors in the sciences; and developing mathematical models using a combination of spreadsheets and general purpose modeling applications that incorporated measured quantities to make predictions. The technology team, the engineering team and the mathematics team all had areas of focus described elsewhere (*1*).

Two cohort groups were sponsored, each participating for 2 consecutive years with an overlap in the second year. For each cohort the first year was considered as a Capacity Building year. The program began with a preparation activity involving a summer course for K-12 STEM teachers, followed by a kickoff meeting as the school year begins, a set of IT intensive STEM area workshops for students during the year and seminar meetings near the end of the fall and winter semesters. The year concluded with real-world field-based experiences during the following summer and opportunities for students to work directly with IT and STEM professionals and see examples of real-world workplace applications. By improving the IT/STEM readiness of participating students through capacity building activities and field-based experiences, all students were prepared to undertake the work of designing IT-intensive projects within the context of STEM, work that begins during the program explorations and continues into the second year, considered as Design year.

The global temperature model explained in this chapter was further developed and improved when teaching Environmental Chemistry to a class of environmental studies graduate students as a regular course taught in the Department of Natural Sciences at the University of Michigan-Dearborn. These later developments produced the model that is mostly described here. Further modifications were later made to account for the average temperatures of neighboring planets Venus and Mars. In addition to Environmental Chemistry, the model continues to find application in Physical Chemistry II where it is presented in conjunction with spectroscopy. This model provides a fairly realistic yet simplistic physical interpretation of how energy from the sun is obtained by the Earth and subsequently trapped by atmospheric gases. Thus the model consists of 2 parts. First, the Earth without atmosphere absorbs radiant energy from the sun and expels this radiant heat upward towards space. The second part of the model accounts for the recapture of select wavelengths of that radiant heat by those atmospheric gases that are infrared-active. The vibrational energy of the IR active molecules is transferred by collision with the more common atmospheric gases such as nitrogen and oxygen.

Climate change is a complex and poorly understood concept that is difficult to ignore when considering the average global temperature. Due to the slow and unpredictable regional oscillations in climate there is difficulty in persuading the public of the severity of anthropogenic contributions to the atmosphere. And, to make matters worse, policy-makers and media outlets continue to assert the uncertainty in climate science (2). As a result, US policy has tended to forego legislation supporting any significant long-term climate planning. To improve the public perception of the sensitivity of global temperatures to human utilization of fossil fuels, broader understanding of how IR active molecules connect with atmospheric thermal energy capture is essential. We present a fairly simple model based on spectroscopic and thermodynamic principles that implicate rising atmospheric CO_2 levels, or any other "greenhouse" gas, result in rising global average temperature.

What makes this approach useful in a service model environment is the ease of understanding of the model and the obvious implications from a first principle and mathematical understanding of science. The model is built using an Excel spreadsheet, ensuring that virtually anyone with a PC or mobile device can apply and modify the model. The model gives students a sense of what each IR active gas contributes to temperature as well as the overall influence that the atmosphere has on climate. Some of the failures of the model include the complete lack of the dynamical processes important to regional climate, contribution of the oceans and the lack of feedback processes. However, an ingenious student could include other factors, especially, if they would like to account for a particular impact, such as including other polluting gases such as chlorofluorocarbons.

The discussion of climate change often includes a description of an increasing global human population along with the increasing food and energy demands required to sustain that population. These increases have resulted in significant output of by-product gases such as carbon dioxide and methane. The consumption of fossil fuels as a primary energy source over the past 130 years has released vast amounts of previously stored carbon in the form of carbon dioxide. The increase in anthropogenic CO_2 has been directly observed through measurements taken at the Mauna Loa Observatory in Hawaii begun under the supervision of Charles D. Keeling of the Scripps Institution of Oceanography at the University of California – San Diego in 1958 (*3*). The Keeling Curve (*4*) not only shows the rise in the concentration of CO_2 in the atmosphere but also shows how the rate of change of CO_2 concentration is increasing. And, while the concentration unit of parts per million (ppm) might sound like a ridiculously small number to the layman, useful examples can be provided to help students understand the implication of such concentrations.

For example, it is useful to indicate to students that CO_2 levels of 5000 to 10000 ppm are considered toxic to humans. From the Keeling curve it is fairly easy to see that by 2100 the global CO_2 levels will have reached 1000 ppm. This is just 1/5 of the concentration of CO_2 considered dangerous to humans. Other examples, not directly related to CO_2 can also be given such as the lethal level of the related gas carbon monoxide. Carbon monoxide is typically produced during inefficient combustion and usually present with fossil fuel combustion. This gas is consider toxic around 1600 ppm. Alternatively, gases such as chlorine become lethal at concentrations as low as 10 ppm (*5*).

One of the most immediate impacts associated with rising average global temperatures are the rising ocean levels. Since 1900, the cumulative contribution to sea level from polar glacial melt and thermal expansion has resulted in a sea level rise of 1.7 mm per year and the rate of sea level rise appears to be increasing as from 1993 through 2010 the rate has been measured to be 3.2 mm per year (*6*). While the rates of mm per year may not seem like much, over the last 110 years the ocean level increase is 19.0 cm. The connection of the increasing rates of rising temperature to increasing rates of rising oceans create some awful predictions for the large populations that live near the ocean shores. Other consequences of rising CO_2 levels, include an increase in pH levels of the oceans due to CO_2's acidic nature, loss of glacier and polar environments and related habitat and an increase in the severity and length storm patterns to name a few (*7*).

Even with the long trending evidence of climate change, there is a large group of Americans who often cite short-term weather predictions as a failure of the climate science. As a result, it is not uncommon to hear about climate change as either too complex or simply as an unpredictable event that may be occurring for reasons that are not anthropogenic. What makes the weather unpredictable, however, is the difficulty in predicting dynamical effects of such a large system on small regions as in the case of an individual's local environment. A specific weather forecast for a small region of observation tends to be a very challenging objective for a meteorologist presenting to a viewing audience encompassing millions, however, the prediction for a large viewing region as a whole tends to be fairly accurate. Climate becomes fairly complex when dealing with the details of daily events, however when working with the averages it is quite accurate.

As we will soon see, a minimum discussion of climate requires knowledge of blackbody radiation, vibrational spectroscopy, selection rules and the understanding that vibrational kinetic energy can be transferred to translational kinetic energy via collisions. The average individual is aware of the sun as the origin of our climate, however rarely is it understood how the atmosphere truly participates in global thermal stability. If the Earth were devoid of an atmosphere, the surface facing the sun would obviously be warm, however, the temperature of the side not facing the sun would fall off dramatically. To account for how the atmosphere retains heat and then quantitatively estimate an average global temperature we need to account how the atmospheric gases trap heat. Providing a simple but accurate description of how the Earth receives incoming energy from the sun and traps some of the heat returning to space is the primary goal of this work.

Group Approach To Developing the Model

Initially, the various high school groups participating in the development of the model had little understanding of climate or climate change. Both teachers and students were primarily educated on climate by the media, documentary television and perhaps limited selections from general science textbooks. Few, if any, had looked into the matter beyond reading an article online. However, when tested, virtually all of the students believe that humans are having a long-term negative impact on the environment and humans are responsible for an increase in the average temperature. These students also believe that rising levels of CO_2 and global temperatures are beyond their control. Students and teachers alike, had only a limited understanding or grasp of the mechanism by which humans are impacting climate. Typical understanding of climate change was limited to the student's local ecosystem and rarely included a broader understanding to regions such as polar or mountainous environments. A common, recurring theme indicated that the summers "seemed" warmer and dryer whereas the winters "seemed" milder and dryer.

In addition to the instructor guiding the students through the various components of the model, environmental studies graduate students were also available to answer questions and assist wherever needed. The graduate students were selected from the Environmental Science Graduate Program of the University of Michigan- Dearborn and came with strong background in quantitative measurement and experience in teaching. However, even the graduate students possessed limited understanding of the relation between select atmospheric gases and global climate. Four short lectures provided the students with the mathematical description of each part of the model. The first lecture took about 10 minutes and consisted of the description of blackbody radiation. During the lecture plots of the radiation curves at different temperatures are shown along with the blackbody spectrum of the Earth. A model curve at 300 K is compared with the Earth's spectrum to identify depressions associated with certain atmospheric gases such as water and CO_2. Following this first lecture, another pair of 15-minute lectures introduced students to the concepts of vibrational absorption and IR active modes. These high school students were familiar with the chemical bonds and with the concept of the dipole moment.

What they were unaware of, not uncommon to be lacking from high school chemistry, is the selection rule requiring a variation of charge through a vibrational mode to absorb/emit radiation. This latter topic is something the students tend to accept rather than fully understand. The last major topic was a 10-minute lecture on heat capacities. These 4 topics were considered separately and at the end of the lectures students were given the schematic shown in Figure 1 connecting equations described in lecture.

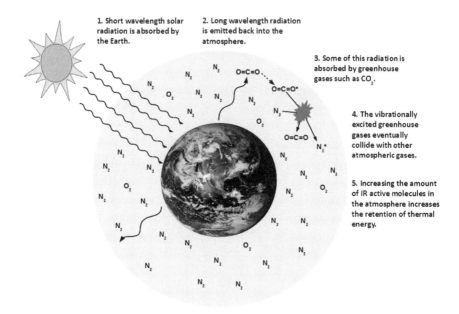

Figure 1. Schematic representation of how CO_2 catches IR radiation heading towards space and transfers this heat to atmospheric gases such as N_2 or O_2.

Students were given roughly 30 minutes to work with the concepts and assemble the components in a manner that produce a sensible prediction for at least one of the vibrational modes of CO_2. As these students attempted to put the parts together, they asked the graduate students and the instructor questions which help them work towards what the author considers a final correct result. The result is described by equation 9 or a change in temperature that corresponds to the retention of heat from the one vibrational mode of an IR absorbing gas.

Climate Models

Understanding and predicting the Earth's climate has long been a challenge in science. One of the first links of carbon dioxide to global climate came with the French mathematician Joseph Fourier in 1824, who by using a simple energy balance approach estimated that the Earth should actually be colder than it is. Fourier came to the conclusion that the Earth's atmosphere acts like the glass in a greenhouse, preventing some radiation from escaping (8). Later John Tyndall used infrared radiation to show that transparent gases such as water and CO_2 absorb IR radiation and as a result trap heat from the absorbed radiation.

The first climate models to indicate that atmospheric CO_2 could have a significant impact on climate were published in the 1960s (9). These models indicate that increased amounts of CO_2 in the atmosphere would alter the absorption and emission of infrared radiation propagating through the atmosphere. Climate forcing, or the result of changing the amount of IR active molecules in the atmosphere, could occur with relatively small changes. For example, the radiation capture by CO_2 would lead to a slight elevation in temperature, the result of which would increase the amount of water vapor. Since water vapor is also IR active, the water vapor would add to the retention of the IR radiation and thus the CO_2 would be responsible for the initiation of feedback. While these early models were relatively simple they provided the first indication that climate could be sensitive to relatively small changes.

Modern climate models range from the relatively simple energy balance approaches to very sophisticated multicomponent interactions requiring complex differential equation based flow dynamics to account for regional and global energy flux, ocean currents and biological life cycles. Looking to the simplest type of model, the zero-dimensional energy balance model (0D-EBM) equates energy from the sun absorbed by the Earth with thermal energy reflected back into space. A 0D-EBM typically results in a single quantity such as the global average temperature but provides an interpretation of interaction between sun and climate. Models of higher order estimate average temperatures and produce meteorological variables over regional areas, usually by latitude. These models can account for the impact of ocean, topography of land and the location and expanse of ice formations. In the Earth model of intermediate complexity (EMIC) all of the fundamental components of the global climate system are included such as atmosphere, ocean, ice, and land masses including soil and vegetation. These models account for time scales from one season to millions of years (10). Another set of sophisticated models are the general circulation models (GCMs). These models depict climate on a global 3 dimensional grid with resolutions of 250 to 600 km horizontal relative to the surface of the Earth and 10 to 20 km vertically through the atmosphere. GCMs account for global convection and feedback processes including those biological and chemical in nature and those processes that influence absorption and emission of radiation to include feedback associated with anthropogenic CO_2 and other radiation absorbing or reflecting pollution. Regional circulation models (RCMs) are run at higher resolution than the GCMs and are thus able to better represent mesoscale dynamics and thermodynamics such as cloud formation, storm fronts, and lake-effect snow. As small scale model of the climate, an RCM can receive lateral boundary conditions from the relatively-coarse resolution output of a GCM (11).

With the exception of the 0D-EBM, the above mentioned modelling methods require a deep understanding of both mathematics, physics and chemistry to truly understand how they work. As a result they are not likely useful to teaching the basics of climate to high school students or college students in an environmental chemistry course. The 0D-EBM serves as the better candidate when understanding the impact of solar radiation and thermal dissipation. There are variations to the 0D-EBM and many these variations insert semi-empirical parameters to account for heat retention by the IR active gases or reflective aerosols. The model presented in this work provides a mechanism detailing the process by which select IR active molecules capture spaceward bound radiation convert that electromagnetic radiation to thermal energy.

The Model: Zero-Dimensional Energy Balance Model Including IR Active Molecules

There are a variety of models and simulations connected with climate. Many of these models use flow dynamics and feedback mechanisms to attempt an accurate and realistic model of the climate. Several simulations that are based on various climate models of varying complexity can be found elsewhere (12). While these models work well, they are very broad and require significant understanding or some serious reading of research articles to understand how the simulations work. Many of these simulators utilize graphical interfaces that reduce input to simply dragging cursors around to increase or decrease various factors. These models leave the user with a "black box" feel for the resulting output. During the early stages of the IT/STEM workshops, a group of high school teachers used a prebuilt climate model created with STELLA (13). The group found it difficult to connect the model with what is occurring at the molecular level, a problem typical of many models that utilize units such as global gross tonnage output or radiative absorption of CO_2 (14). It is this non-molecular approach that created an incentive to build a climate model based upon molecular interactions. A model that considers the relative concentration of CO_2 and other gases, near the surface of the Earth. What makes this model useful to student learning is the building up approach from basic molecular principles in a simplified but realistic approach.

Our understanding of climate begins with a description of the Earth's energy balance excluding atmospheric gases. This understanding is constructed from the idea that the total energy entering the Earth as equivalent to the energy radiated back into space. A formula based on the conservation of energy for any planetary system can be reduced to a few terms giving the temperature and is shown by the energy balance equation (EBE)

$$T = \left(\frac{(1-A)F_s}{4 S_b}\right)^{1/4} \qquad 1$$

Where F_s is the energy received from the sun called the solar flux, 1368 Wm^{-2}, A corresponds to the solar energy immediately reflected by surfaces such as clouds or desert sand called the albedo, 0.31 and S_b is a term that corresponds to the thermal energy given off by the planetary body as direct result of its temperature and is called the Stefan-Boltzmann constant, 5.67×10^{-8} $Wm^{-2}K^{-4}$. This last term is likely the most challenging for students as it relates to something called blackbody radiation. All objects produce radiation as a result of their having thermal energy. Thermal energy corresponds to molecular translation, rotation and vibration. When the molecules are fixed in a solid the thermal energy is limited to the vibrational motion of the atoms. These oscillating charged particles result in the emission of electromagnetic radiation of frequencies equivalent to the frequencies of the oscillating charges. The higher the temperature an object, the more common the higher frequency oscillations and the greater the number of higher frequency photons. All objects, regardless of chemical nature, emit an electromagnetic energy that is characteristic of only the temperature. Thus whether it is an entire planet or just some part of the planet like a body of water, the emission spectrum at a given temperature has the same characteristic shape and thus a predictable emission energy. The derivation of the equation above can be found elsewhere (*15*).

Using commonly accepted values, the Earth's energy balance corresponding to the radiant energy of the sun interacting with the Earth and excluding atmospheric gases results in a temperature of 254 K (-19 °C). The average temperature of the Earth is 288 K (15 °C) and therefore the energy balance formula underestimates the Earth's average temperature by 34 K. However, if we apply this formula to a system without an atmosphere, such as the moon, then the average temperature is overestimated by 4 K. When we apply this formula to a planet with a thin atmosphere such as Mars, the temperature is overestimated by 5 °C. The EBE works well for systems with little or no atmosphere. In the case of the moon or Mars, the lack of atmosphere allows radiant energy to escape. However with a significant atmosphere such as the Earth's, the observed temperature is greater because the atmosphere retains heat. To improve the EBE, one must account for atmospheric gases.

Our approach to modifying the EBE is to effectively add, or account for, captured thermal energy in the atmosphere. Much of the planet's thermal energy is emitted back into space and only a small amount is collected by certain atmospheric gases. Select frequencies of the returning radiation are captured by IR active molecules as vibrational energy. The captured vibrational energy is converted into thermal energy by collision with atmospheric nitrogen and oxygen. This is the basic mechanism for how the "greenhouse gases" trap heat in the atmosphere. Unfortunately, such a description is often discarded as many layman are unfamiliar with concepts such as heat capacities or molecular vibration.

For the first attempt at an atmospheric IR adsorption model we limit consideration to 4 topics learned in general chemistry. The blackbody spectrum of the Earth, relative concentrations of gases in the atmosphere, the vibrations associated with polyatomic molecules and finally, the heat capacity of gaseous molecules in the atmosphere. Clearly, these topics are well understood by second semester physical chemistry students. And, depending on the level of the student, some of these topics can be summarized at a fairly rapid pace.

Carbon dioxide is a triatomic molecule with no permanent dipole moment. However, the molecule has 4 vibrational modes. Three of these modes have a symmetry that cause a non-uniform displacement in charge through the motion of the vibration. The frequency of these particular oscillations correlate directly with the frequency of light that can be absorbed by the molecule. The magnitude of the variation in the charge dictates the intensity or the amount of light that is absorbed by the vibrational mode. One of the first most significant simplifications involves identifying the IR active vibrational modes and then computing the energy of a photon of light associated with that mode. The model assumes that all of the energy is kinetically transferred to the solvent atmospheric gases, nitrogen and oxygen. The model also excludes contributions from the rotational spectra.

For example, CO_2 has 4 vibrational modes. A pair of degenerate IR active modes at 667 cm^{-1}, another IR active mode at 2349 cm^{-1} and a non-IR active mode at 1380 cm^{-1}. Rather than exclude the non-IR mode, we simply multiply this mode by 0, indicating that it has no intensity. The degenerate IR active modes are counted twice. To calculate the energy absorbed by each CO_2 molecule at 667 cm^{-1} we compute

$$E_{CO_2}^{667\ cm^{-1}} = h\ [J\ s]\ c\ [cm\ s^{-1}]\ 667\ [cm^{-1}] = 1.326 \times 10^{-20} J \qquad 2$$

Where h is Planck's constant and c is the speed of light. If we assume every available CO_2 molecule absorbs this energy, then the total energy absorbed by some arbitrary unit of atmosphere depends on the concentration of CO_2. The energy of each photon becomes vibrational energy. This energy can be released either by emitting a photon of equivalent energy or via collision with the more common nitrogen and oxygen present in the atmosphere. Assuming the energy is transferred we have

$$E_{CO_2}^{vib} = E^{air} \qquad 3$$

Where E^{air} is determined from chemical calorimetry as

$$E^{N_2\ and\ O_2} = m C_s^{air} \Delta T \qquad 4$$

Recently, atmospheric CO_2 levels measured at Mauna Loa reached a new milestone of 400 ppm. We consider the radiation absorbed by the CO_2 molecules in 1 atm of atmosphere which is simply 400 molecules of CO_2 for every 10^6 molecules of air. Beginning with the formula:

$$E^{Transfer} = m C_s^{air} \Delta T \qquad 5$$

So for CO_2 at 400 ppm we can write

$$400 * E_{CO_2}^{vib} = m C_s^{air} \Delta T \qquad 6$$

And for every 400 CO_2 molecules the mass of atmospheric air can be written as

$$m = \frac{10^6}{6.022 \times 10^{23}} * 28.97\ g\ mol^{-1} \qquad 7$$

Where the 28.97 g mol^{-1} corresponds to the average molar mass of air. We can write a formula for the increase in temperature of air due to a single vibrational mode of an IR active gas as

$$\Delta T = \frac{C_{gas}^{ppm} * E_{mode}^{vib} * N_A}{10^6 * M_{air} * c_{air}} \qquad 8$$

Where C_{gas}^{ppm} corresponds to the concentration of the gas, N_A is Avogadro's number, M_{air} is the molar mass of air, and c_{air} is the heat capactity of air. For the 667 cm^{-1} IR mode of CO_2 at 400 ppm we have:

$$\Delta T = \frac{400 * 1.326 \times 10^{-20} J\ CO_2^{-1} * 6.022 \times 10^{23} CO_2\ mol^{-1}}{10^6 * 28.97\ g\ mol^{-1} * 1.070\ J\ g^{-1}\ K^{-1}} \qquad 9$$

Which yields

$$\Delta T = 0.103\ K \qquad 10$$

This ΔT is the contribution from the one vibrational mode of CO_2 to the atmosphere. Doubling the number because this mode is degenerate and then performing a similar calculation for the other IR active mode at 2349 cm^{-1} and adding the resulting 3 numbers corresponds to a temperature increase of 0.57 K. This temperature adjustment suggests that at 400 ppm, CO_2 accounts for a 1.6% contribution to the temperature of the atmosphere. At 280 ppm CO_2, roughly preindustrial levels, the temperature contribution is 1.1%. This value seems sensible. In fact, if we compare our value with predicted changes from other models, the range is typically from 1 °C to 3 °C depending on the model.

Following the example of CO_2, other IR active gases that are significant in concentration in the atmosphere to be considered include: O_3 (10 ppm), CH_4 (1.79 ppm), N_2O (0.31 ppm) and the most significant greenhouse gas H_2O (10000 ppm) where water averages an atmospheric concentration of 1%. When the temperature contributions of each molecule added together the model's total temperature contribution from the IR active gases is 35.4 °C. Adding this value the temperature produced using the EBE brings the final average global temperature to 289.4 K which is only 1 °C below the observed average. Table 1 contains a spreadsheet indicating the temperature contributions from each gas included in this model.

Venus and Mars

Application of the EBE to Venus and Mars produce average temperatures of 261 K and 215 K, respectively. The average observable temperatures of these neighboring planets are 730 K and 212 K, respectively. The reason for the large discrepancy between the EBE and the observed average temperature of Venus is a result of the planet's atmospheric conditions. Venus has an atmospheric pressure that is 92 times that of the Earth's whereas the Martian atmosphere is only 0.6% of the Earth's atmospheric pressure. Venus has a mass and thus gravitational pull similar to Earth whereas Mars is much smaller and less capable of retaining atmospheric gases. Venus has a total atmospheric mass of 4.8x10^{20} kg whereas Earth has an atmospheric mass of 5.1x10^{18} kg. Also of interest is the composition of the atmospheres. Both Mars and Venus have a similar concentration of CO_2. The atmosphere of Venus is 96.5% CO_2 and 3.5% N_2 and the composition of Martian atmosphere is 95.3% CO_2 and 2.7% N_2 (16). The differences in the atmospheres of the 3 planets have created very unique conditions.

Table 1. Spreadsheet Table of the Thermal Contributions by Common IR Absorbing Atmospheric Gases

Earth	Conc [ppm]	IR Mode [cm^{-1}]	IR active	Molar Energy [J mol^{-1}]	ΔT [K]	Total Contrib [K]	Contrib %
CO_2	400	667	1	1.33E-20	0.103		
		667	1	1.33E-20	0.103		
		1380	0	2.74E-20	0.000		
		2349	1	4.67E-20	0.363	0.57	1.61%
O_3	10	1103	1	2.19E-20	0.004		
		701	1	1.39E-20	0.003		
		1042	1	2.07E-20	0.004	0.01	3.11%
CH_4	1.7	1360	1	2.70E-20	0.001		
		1362	1	2.71E-20	0.001		
		1363	1	2.71E-20	0.001		
		1450	1	2.88E-20	0.001		
		1452	1	2.89E-20	0.001		
		3206	1	6.37E-20	0.002		
		3209	1	6.38E-20	0.002		
		3213	1	6.39E-20	0.002		
		3314	1	6.59E-20	0.002	0.01	0.04%
N_2O	0.31	596	1	1.19E-20	0.000		
		1298	1	2.58E-20	0.005		
		2282	1	4.54E-20	0.009	0.01	0.04%
H_2O	10000	3657	1	7.27E-20	14.122		
		1595	1	3.17E-20	6.159		
		3756	1	7.47E-20	14.505	34.79	98.29%
					ΔT [K]	35.4	
					T_{EBE} [K]	254.0	
					$T_{ODEBMIR}$ [K]	289.4	%diff.
					T_{obs} [K]	288	0.3%

Application of equation 9 to the 667 cm^{-1} vibrational mode of CO_2, along with the other appropriate terms, produce temperatures of 201 K for Venus and 91 K for Mars. According to the model just 1 vibrational mode of CO_2 alone contributes more to the temperature than is observed. There are several reasons for this disagreement and this section considers the 2 most significant. First, the atmospheres of both planets contain a much higher concentration of CO_2 than does the Earth. The irradiance or energy density of the blackbody radiation being emitted by a planet limits the amount of radiation available for absorption. In the case of Earth it is reasonable to assume that there are more photons at the frequency corresponding to the vibrational modes of the CO_2 molecules than there are molecules. In the case of Venus and Mars, however, there are significantly more CO_2 molecules than photons of a particular frequency.

The other significant factor is the relative atmospheric pressure of the planets. Mars has an atmospheric pressure of 626 Pa, much less than the Earth's. The low pressure allows the EBE to work fairly well in describing the planets average temperature. Venus has an atmospheric pressure much higher than the Earth's and therefore Venus retains more heat. The importance of the higher atmospheric pressure in heat retention can be observed by the diurnal temperature swing. Venus shows little variation from day to night whereas Mars has large swings in temperature.

To account for the effects of the amount of IR absorbing gas present relative to the total atmospheric pressure we introduce the following term

$$\lambda_{limit} = 0.066 \, Ln\left(\frac{\frac{P_{atmosphere}}{101325 \, Pa}}{X_{gas}}\right) + 0.3339 \qquad 11$$

Where $P_{atmosphere}$ is the total atmospheric pressure of the planet in units of Pascal, 101325 Pa is a reference quantity used to cancel units and X_{gas} is fractional composition of the IR active gas of interest. The resulting λ_{limit} is a dimensionless correction factor that can be applied to each of the 3 planets and all of the IR active gases.

The inclusion of λ_{limit} changes equation 8 to

$$\Delta T = \frac{C_{gas}^{ppm} * E_{mode}^{vib} * N_A * \lambda_{limit}}{10^6 * M_{air} * c_{air}} \qquad 12$$

And the result of equation 12 with values for the 667 cm^{-1} IR mode of CO_2 at 400 ppm for Earth remains virtually unchanged with a value of 0.103 K. When applied to conditions on Venus, however, the temperature contribution from the 667 cm^{-1} of CO_2 yields

$$\frac{965000 * 1.326x10^{-20} J \, CO_2^{-1} * 6.022x10^{23} * CO_2 \, mol^{-1} * 0.391}{10^6 * 44.01 \, g \, mol^{-1} * 1.36 \, J \, g^{-1} \, K^{-1}} = 81.6 \, K \qquad 13$$

Where the concentration of CO_2 is taken as 96.5% of 10^6, the molar mass and the heat capacity of the atmosphere is effectively that for CO_2 or 44.01 g mol^{-1} and 1.36 J g^{-1} K^{-1} (17), respectively. Considering the remaining CO_2 vibrational modes and other IR active gases, the model obtains a total average temperature for Venus of 711.3 K. This value is 450.7 K above the temperature predicted by the EBE and only 19 K above the observed temperature. The model also performs well for Mars predicting an average temperature of 216.5 K. A value that is 4 K above the observed temperature. It is important to note that the EBE value is already 3 K above the observed temperature.

Table 2. Spreadsheet Table of the Thermal Contributions by Common IR Absorbing Atmospheric Gases on the Planets Venus

Venus	Conc [ppm]	IR Mode [cm⁻¹]	λ_{limit}	Molar Energy [J mol⁻¹]	$\Delta T[K]$	Total Contrib [K]	Contrib %
CO_2	965000	667	0.634	1.33E-20	81.591		
		667	0.634	1.33E-20	81.591		
		1380	0.000	2.74E-20	0.000		
		2349	0.634	4.67E-20	287.343	450.52	99.9%
SO_2	150	507	1.213	1.01E-20	0.018		
		1145	1.213	2.28E-20	0.042		
		1334	1.213	2.65E-20	0.049	0.11	0.0%
H_2O	20	3657	1.346	7.27E-20	0.020		
		1595	1.346	3.17E-20	0.009		
		3756	1.346	7.47E-20	0.020	0.05	0.0%
				ΔT [K]	450.7		
				$T_{EBE}[K]$	260.6		
				$T_{ODEBMIR}[K]$	711.3		%diff.
				$T_{obs}[K]$	730		2%

Inclusion of this correction factor for atmospheric pressure and relative amount of CO_2 to the various IR active gases of each planet are presented in Table 2 and show a relatively good agreement. And while this correction factor does not account for all of the nuances associated with the IR active gases and the corresponding heat exchange, it does allow a reasonable computation of the average planetary temperatures without adding too much to the complexity of the model.

Range of Students

At a minimum, this model requires a reasonable understanding of vibrational spectra, heat capacities and concentrations. Thus the applicability of the model is limited to those students with at least some high school chemistry and/or physics. We applied this model to senior high school students who had comparatively significant backgrounds and reasonable interests in science. These students were also somewhat familiar with the aspects of climate change and typically believed that anthropogenic sources of greenhouse gases are perturbing the average global temperature.

Table 3. Spreadsheet Table of the Thermal Contributions by Common IR Absorbing Atmospheric Gases on the Planets Mars

Mars	Conc [ppm]	IR Mode [cm^{-1}]	λ_{limit}	Molar Energy [J mol^{-1}]	$\Delta T[K]$	Total Contrib [K]	Contrib %
CO_2	960000	667	0.001	1.33E-20	0.178		
		667	0.001	1.33E-20	0.178		
		1380	0.000	2.74E-20	0.000		
		2349	0.001	4.67E-20	0.628	0.98	92.2%
H_2O	210	3657	0.557	7.27E-20	0.013		
		1595	0.557	3.17E-20	0.006		
		3756	0.557	7.47E-20	0.013	0.03	3.0%
NO	100	1955	0.606	3.89E-20	0.038	0.05	4.8%
					$\Delta T[K]$	1.1	
					$T_{EBE}[K]$	215.5	
					$T_{0DEBMIR}[K]$	216.5	%diff.
					$T_{obs}[K]$	212	-3%

Assessment

The purpose of the modeling portion of the program was to develop interest in computer modeling and to enable the students to develop models for an independent science based project of their choosing. Limited assessment associated directly with this 0D-EBM-IR was performed with the IT/STEM students. Rather assessment activities were related to computer modeling as a whole (1). The 1st cohort worked with the generalizable modeling application STELLA where the 0D-EBM was obtained from a library of sample systems. Students were able to modify the STELLA model by changing variables and watch animated outcomes as a result of their changes. The students understood the consequences of making changes to the amount of greenhouse gases however it was clear that they really did not understand the mechanisms of the model. For the 2nd cohort of students, we included a description of energy balance and presented concepts of how atmospheric CO_2 absorbs the emitted IR and via collisions transfer that energy to the atmosphere as heat. Following this approach the students acquired a clearer grasp for inner workings of the 0D-EBM-IR as well as global climate. The idea of the 0D-EBM-IR model described in this work came to fruition following the experience with the 1st cohort. It was the 2nd cohort of students that developed the ideas closer to what is presented here. As a result, the 2nd cohort's experience with modeling was clearly more developed.

Environmental Chemistry

This model was later applied to students taking a graduate course in environmental chemistry. While these students lacked a strong background in some of the topics needed to understand this model, the material is available in many environmental chemistry textbooks. Concepts such as the IR active mode are needed to describe infrared spectroscopy, blackbody radiation is necessary to understand climate and thermal energy. Many environmental textbooks even use topics connected with calorimetry and thermal chemistry. Even though this approach is mathematical, the model fits well with many environmental studies courses and leaves students with solid understanding of the molecular nature of climate.

Conclusion

Most areas of science and engineering require some form of computer modeling to make valuable predictions that assist in providing direction or explanation. Yet, the teaching of modeling and modeling techniques in the classroom is still a subject that high school or college students rarely see. In this chapter we propose a new model, the zero-dimensional energy balance model with corrections of IR active atmospheric gases, useful in predicting the average global temperature. The model was originally developed to be presented to high school students, however, the model makes for valuable lessons with college students. We have successfully applied this model to the various student groups and indicate that the students obtained a better understanding for how climate works. Students are also equipped with an understanding of how anthropogenic CO_2 may be having an impact on the Earth's climate.

In science, even the most ardent experimentalists ultimately requires models to describe, explain and categorize results. Models are the way humans conceptualize what we perceive around us and it is with these models that we can explain how something works. There are people who argue that "computer" modeling, in particular, is problematic and should always be avoided when experimental data is available. However, there is little experimental data that does not include extrapolation or interpolation which is in fact just a model for where there is no data. Furthermore, ignoring computer models is short-sighted. Much like the development of experimental techniques in the last 200 years, development in computer technology and computer algorithms in the next century may even exceed experiment.

References

1. Duran, M.; Hoft, M.; Lawson, D. B.; Medjahed, B.; Orady, E. A. Urban high school students' IT/STEM learning: Findings from a collaborative inquiry- and design-based afterschool program. *J. Sci. Educ. Technol.* **2014**, *23*, 116–137.
2. Oreskes, N. The Scientific Consensus on Climate Change. *Science* **2004**, *306*, 1686.

3. Keeling, C. D. The Concentration and Isotopic Abundances of Carbon Dioxide in the Atmosphere. *Tellus* **1960**, *12*, 200–203.
4. *Trends in Atmospheric Carbon Dioxide*; http://www.esrl.noaa.gov/gmd/ccgg/trends/ gives the latest CO_2 levels taken from Mauna Loa (accessed Oct. 14, 2014).
5. *CRC Handbook of Chemistry and Physics*, 89th ed.; Lide, D. R., Ed.; CRC Press: Boca Raton, FL, 2003; Section 3, p 476.
6. Church, J. A.; Clark, P. U.; Cazenave, A.; Gregory, J. M.; Jevrejeva, S.; Levermann, A.; Merrifield, M. A.; Milne, G. A.; Nerem, R. S.; Nunn, P. D.; Payne, A. J.; Pfeffer, W. T.; Stammer, D.; Unnikrishnan, A. S. Sea Level Change. In *Climate Change 2013: The Physical Science Basis. Contribution of Working Group I to the Fifth Assessment Report of the Intergovernmental Panel on Climate Change*; Cambridge University Press: New York, 2003; pp 1137–1216.
7. *The Discovery of Global Warming: The Carbon Dioxide Greenhouse Effect*; URL http://www.aip.org/history/climate/co2.htm (accessed Feb. 19, 2014).
8. Cowie, J. *Climate Change: Biological and Human Aspects*; Cambridge University Press: New York, 2007; p 3.
9. Möller, F. On the Influence of Changes in the CO_2 Concentration in Air on the Radiation Balance of the Earth's Surface and on the Climate. *J. Geophys. Res.* **1963**, *68*, 3877–3886.
10. Petoukhov, V.; Claussen, M.; Berger, A.; Crucifix, M.; Eby, M.; Eliseev, A. V.; Fichefet, T.; Ganopolski, A.; Goosse, H.; Kamenkovich, I.; Mokhoy, I. I.; Montoya, M.; Mysak, V.; Sokolov, A.; Stone, P.; Wang, Z.; Weaver, A. J. EMIC Intercomparison Project (EMIP–CO_2): comparative analysis of EMIC simulations of climate, and of equilibrium and transient responses to atmospheric CO_2 doubling. *Clim. Dyn.* **2005**, *25*, 365–385.
11. McGuffie, K.; Henderson-Sellers, A. *A Climate Modelling Primer*, 3rd ed.; John Wiley & Sons: New York, 2013; p 253.
12. *Climate modeling resouces*; http://www.inscc.utah.edu/~strong/modeling.html (accessed Oct. 14, 2014).
13. *isee sytems*; http://www.iseesystems.com (accessed Oct. 14, 2014).
14. This is not a criticism of the simulations, just a note of one aspect of teaching a complex but likely more accurate, approach.
15. See, for example: Girard, J. E. *Principles of Environmental Chemistry*, 3rd ed.; Jones & Bartlett Learning: Burlington, MA, 2014; pp 80.
16. *NASA Lunar and Planetary Science*; http://nssdc.gsfc.nasa.gov/planetary/ (accessed Nov. 1, 2014).
17. Ernst, G.; Maurer, G.; Wiederuh, E. Flow calorimeter for the accurate determination at high pressure; results for carbon dioxide. *J. Chem. Thermodyn.* **1989**, *21*, 53–65.

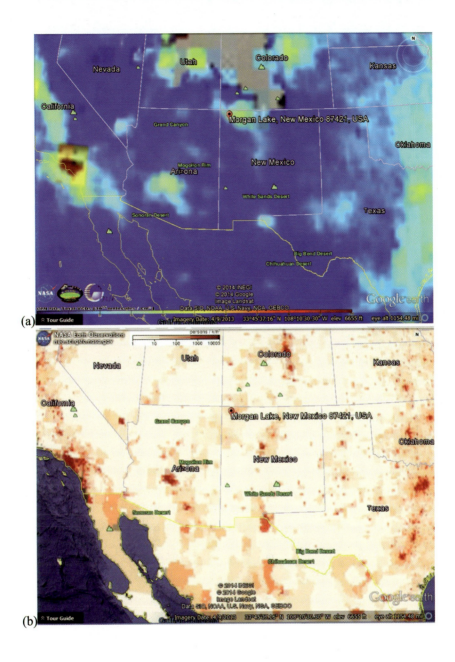

Figure 3. NASA data visualized in Google Earth. (a) Nitrogen dioxide (NO_2) from the Aura satellite and (b) population density distributions over the Southwest United States in January 2014. In both cases, the brighter the color, the higher the concentration. (Chapter 7)

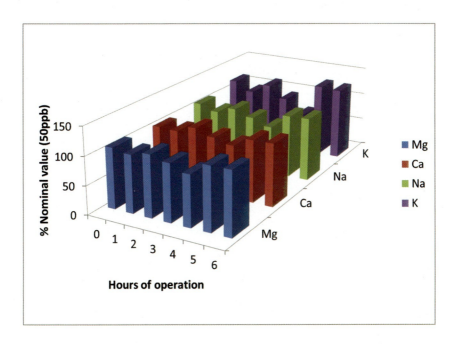

Figure 4. Instrument calibration stability check. This was done with the 50 ppb calibration standard solution. (Chapter 10)

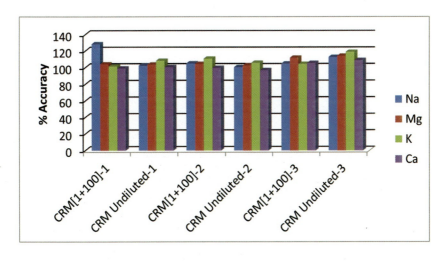

Figure 5. Instrument accuracy using CRM-TMDW-B. (Chapter 10)

Chapter 7

Environmental Justice through Atmospheric Chemistry

Nicole C. Bouvier-Brown*

Department of Chemistry and Biochemistry, Loyola Marymount University,
1 LMU Drive MS 8225, Los Angeles, California 90045
*E-mail: nbouvier@lmu.edu.

The burdens of air pollution are not equally shared among all people. Air quality data, whether extracted from online databases or collected in the field, can be used to demonstrate the patterns of exposure to air pollution. When students explore data trends from recently collected data, their level of intellectual and emotional connection to an environmental justice issue is greater than if they simply read journal articles or case studies. This type of involvement increases awareness of the inequalities and can potentially lead to community action. This chapter outlines three example exercises that highlight the link between air pollution and environmental justice.

Adding the Social Context to Environmental Chemistry

Within Environmental Education, courses typically focus on Environmental Science (ecology, environmental chemistry) or Environmental Studies (environmental ethics, environmental economics) as distinct topics. To understand ecological processes or chemical systems in the environment, one must understand the social context. In this way, the environment becomes a place rich with social, economic, political, and historical contexts that frame the ecosystem within (*1*). For example, the addition of social context can humanize abstract notions about pollution distribution. Putting a human face on environmental issues can then impact decision-making and policy (*2*). Discussing issues of environmental justice is one way to humanize environmental problems. This chapter discusses

© 2014 American Chemical Society

the connection between atmospheric chemistry and environmental justice. Three example exercises are outlined for instructors who would like to address this topic in their courses.

Introduction to Environmental Justice and Air Pollution

Environmental justice awareness in the United States has its roots in the Civil Rights movement of the late 1960s and 1970s. National public awareness grew when, in 1982, protesters in a primarily African-American community in Warren County, North Carolina, tried to prevent the dumping of polychlorinated biphenyl (PCB)-contaminated soil in the nearby landfill. Disparities in environmental pollution burden were then studied throughout the 1980s and '90s, giving credibility to the environmental justice movement (*3*). On February 11, 1994, President Bill Clinton signed Executive Order 12898, "Federal Actions to Address Environmental Justice in Minority Populations and Low-Income Populations" directing federal agencies to address the disproportionately high health and environmental effects of their programs and policies on minority and low-income populations. This order calls on each federal agency to "make achieving environmental justice part of its mission by identifying and addressing, as appropriate, disproportionately high and adverse human health or environmental effects of its programs, policies, and activities…" (*4*). The U.S. Environmental Protection Agency (EPA) chairs this interagency endeavor. The EPA defines environmental justice as "the fair treatment and meaningful involvement of all people regardless of race, color, national origin, or income with respect to the development, implementation, and enforcement of environmental laws, regulations, and policies." This goal "will be achieved when everyone enjoys the same degree of protection from environmental and health hazards and equal access to the decision-making process to have a healthy environment in which to live, learn, and work" (*5*).

Exposure to air pollutants, such as carbon monoxide (CO), nitrogen oxides (NO_x), ozone (O_3), particulate matter (PM), and volatile organic compounds (VOCs), has been related to occurrences of respiratory and cardiovascular diseases as well as hospital admissions ((*6*) *and references therein,* (*7*)). For example, 1999 and 2000 data show that ozone exposure was the strongest predictor of asthma hospitalizations in Phoenix, AZ (*8*). There are also positive correlations between high CO and PM concentrations with low birth weight and prematurity (*9–12*). In addition, school performance decreases in areas with high levels of known respiratory irritants, which, in Southern California, disproportionally affects the African American and Latino children (*13*).

The focus of this chapter is on anthropogenic air pollutants, most of which are a byproduct of the combustion process. CO is a primary pollutant emitted directly from incomplete combustion. NO_x is the sum of nitrogen monoxide (NO) and nitrogen dioxide (NO_2), where NO is directly emitted from sources of combustion and quickly reacts in the atmosphere to become NO_2. NO is the result of the recombination of atmospheric nitrogen (N_2) and oxygen (O_2) that are broken apart in high temperature processes. Ozone (O_3), on the other hand, is a secondary pollutant formed by reacting NO_x and VOCs in the presence of sunlight. Particulate matter (PM) can either be a primary pollutant (e.g., soot) or a product of chemical reactions in the atmosphere, thus becoming a secondary pollutant. Volatile organic compounds (VOCs), gas-phase hydrocarbons, play a role in the formation of ground-level ozone and the growth of secondary particulate matter. In the context of this chapter, the focus is on anthropogenic VOCs, particularly those emitted from fuel and fuel combustion.

Concentrations of air pollutants are correlated with the socioeconomic status of residents. For example, modeled CO and NO_2 concentrations in Birmingham, England were strong predictors of ethnicity and poverty (*14*), and a national study of mean annual outdoor PM showed higher concentrations in socially deprived areas of New Zealand (*15*). These trends are mirrored in the United States (*16*). For example, predicted NO_2 concentrations correlate negatively with household income and positively with poverty in Worcester, MA (*17*), and populations with lower socioeconomic positions were exposed to higher PM in the Northeastern United States (*18*). In the Los Angeles Basin, air pollution is not equally shared among all residents; race alone can explain the risk of air toxin exposure in Southern California (*19*). These studies show that the health burden that this exposure brings is disproportionately shared amongst the residents.

Hands-On Learning

Students can read journal articles or case studies to learn about environmental justice issues related to air pollution, but there is a natural tie-in with respect to service-learning. From an educational perspective, hands-on activities using real data provide students the opportunity to describe trends, construct explanations, and communicate ideas. This inquiry-based approach is essential to learning science (*20, 21*). The informational impact will be much greater to the students if they can personally explore datasets. The academic information also becomes more meaningful because these concepts are directly connected to real life (*22*).

Service-learning also has non-academic benefits. Service-learning that is truly integrated into academic courses allows students to develop a greater awareness of social problems. Students who have ownership over their service-learning projects are more tolerant of others and increase their political engagement (*23*). Warren (*24*) showed that, under instructor supervision, an activity designed to invoke emotional response to environmental injustices can have very positive outcomes. After guided discussions to process these emotions, the students' despair was transformed into actions, including building community gardens, tutoring, and writing letters to mainstream environmental groups (*24*).

The projects discussed here will not only increase our understanding of pollution inequities, but they will also further educate students about the prevalence of this type of social justice. These projects allow students to think about the sources and outcomes of the air pollution distribution in specific communities. Students will learn a direct social application of the science, namely that scientific methods, measurements, and analyses can be used to better our society.

Publically Available Datasets

For air pollutants, like O_3, CO, NO_x, PM, and some VOCs, there are extensive public databases. In the United States, the EPA has a repository of air quality data from over 10,000 ground monitors through the Air Quality System (AQS), and the EPA provides access to the data through various platforms, depending on the requirements of the analysis. There are many websites designed for the general public which are visually appealing and easy to understand (e.g., Air Compare, Air Data, and AIR Now). Other databases provide large quantities of data for researchers and government analysts (e.g., Air Trends, AQS Data Mart, AQS Data Page). The EPA also provides pollutant emission data. For example, the National Emissions Inventory (NEI) database is based on air emission estimates and emission model inputs of hazardous pollutants from known sources provided by state, local, and tribal agencies.

There are also state-specific air quality data. For example, California, a populous state with historic air quality problems and strict regulation practices, provides access to current and historical air quality data through the California Air Resources Board (ARB). Comprehensive raw data from more than 250 monitoring sites (covering 35 air districts) from 1980 – 2011 is available on request (25), but there is extensive online access to filtered data through Air Quality and Meteorological Information System (AQMIS) (26) and iADAM: Air Quality Data Statistics (27). The user can specify the particular time-frame(s), area(s) of California, and pollutant(s) of interest. Like the U.S. EPA, the California ARB also offers emission inventory data, broken down by specific source type and time frame.

Federal agencies have also provided open-access to air quality data derived from satellites. Data from satellite instrumentation have low spatial resolution, but the information often covers the entire Earth. There are many ways to view satellite data. For example, the National Aeronautics and Space Administration (NASA) supports the NASA Earth Observations (NEO) (28), Giovanni (29), and the Socioeconomic Data and Applications Center (SEDAC) (30) websites. These sites, amongst others, collect satellite data and allow users access to maps for visualizing global patterns. Most of the information can also be downloaded for use in other programs. For example, much of the data can be downloaded as Google Earth (.kmz) files.

When relating environmental parameters to environmental justice, there is also a need for co-located demographic information. The most extensive dataset in the United States is the Census; it has been used by many researchers studying air quality and environmental justice (16, 18, 31, 32). Wealth can be measured in median household income, percentage of household living under poverty level, the percentage receiving public assistance, and/or the percentage of single-parent families. Data can also be extracted on age, race, and ethnicity. Some Census data has become more accessible to the public. For example, *The Los Angeles Times* created "Mapping L.A. Neighborhoods" (33) to share compiled demographic statistics for each of the 272 neighborhoods in Los Angeles County. This compilation uses data from the 2000 Census, the Southern California Association of Government, and the Los Angeles Department of City Planning.

Direct Local Sampling

Direct measurements do not rely on extrapolation or estimation of real-time chronic human exposure to air pollutants. When possible, it is important to provide students with hands-on experience of these direct measurements. In this way, students are able to experience analytical methodology – creating standard mixtures, calibrating data, and using instrumentation – and the chemical theory of the application, while also gaining "real world experience". By collecting their own data, students must consider experimental design. The National Science Education Standards emphasize the importance of framing science questions, forming testable hypotheses, and conducting scientific investigations (20). Students are not only learning a direct social application of science, they are also interacting with and learning about their local community.

Ground measurements can be done with expensive instrumentation that have high time-resolution and low detection limits, but that level of sophistication is often unnecessary for environmental justice projects, especially if the project is based on comparison studies or relative changes over time. It is often more important to have a greater number of less-expensive sensors. The most inexpensive ways to detect air pollutants, such as ozone or NO_x, involve simple chemical reactions followed by a colorimetric technique. Test-strips (like the Eco-badge for ozone) and diffusion tubes (like those from Ormantine or Gradko for NO_x) can be purchased from multiple vendors. These techniques can also be replicated using materials in the laboratory. For example, ozone can be detected by reacting it with potassium iodide and a starch indicator, following the Schonbein technique (34). NO_x diffusion tubes can also be made in the laboratory (35). Measuring particulate matter does require instrumentation, but there are some lower-cost options, especially if there is not a need to count particles below ~0.5-1μm (e.g., Dyclos Dc1100, Sharp GP2Y1010AU0F, Dusttrak Laser Photometer from TSI Inc.).

Recently the U.S. EPA has realized the need for inexpensive air sensors that can be used as personal monitors, powered by cellular telephones, or very small sensors that can be bundled together and widely dispersed through "citizen science" (36). These air sensors have amazing capabilities, but most are not yet widely available. For example, the M-pod would cost $300 and measure CO, O_3, NO_2, and total VOCs using metal oxide sensors (37). Additional examples are outlined by Snyder et al. (38).

There are hundreds or thousands of individual trace volatile organic compounds (VOCs) present in the atmosphere at any given time. Not all VOCs have the same environmental impact due to their inherent differences in how effectively they create ozone and/or form particulate matter. In addition, individual VOCs can also have dissimilar health impacts. Thus, it is often important to quantify individual VOCs. Gas chromatography is the best way to separate these gas-phase hydrocarbons, but this makes field work more challenging. While a gas chromatograph (GC) is commonly found in Chemistry departments, they are typically not designed for direct ambient air sampling in the field. Field air samples for VOC analysis can be collected using a container, such as a stainless steel canister, glass flask, or an inert plastic bag. The plastic sample bags are inexpensive, but they can be easily punctured and they have sample artifacts that make "unknown" samples difficult to interpret (*39, 40*). Regardless of container, these air samples still require analyte pre-concentration prior to analysis. The collection and pre-concentration steps can be simultaneously accomplished using a solid adsorbent (*39–41*). Not only is the solid adsorbent easy to handle, but it is also inexpensive; in fact, the adsorbent is at most one-tenth the cost of the equivalent canister. After collection, VOCs are thermally desorbed into the carrier gas flow of a GC (*42*) or extracted from the adsorbent with a solvent and subsequently injected into instrumentation for analysis (*43*). The solvent extraction method is advantageous because it allows for the detection of compounds with high boiling points and it avoids thermal decomposition of any analytes (*44*). It also provides the opportunity for multiple analyses from a single sample (*45*). On the other hand, adding a solvent to extract the analytes dilutes the sample. As a result, longer sampling times are needed to detect ambient-level VOCs. This drawback is out-weighed by the initial equipment cost needed for thermal desorption procedures.

Bouvier-Brown et al. (*46*) developed an inexpensive VOC analysis method using the HayeSep Q solid adsorbent. An additional advantage of this method is that it focuses on VOCs with higher molecular weight and lower polarity. These compounds are typically excluded from traditional air quality studies (and thus governmental databases) due to analytical limitations. These larger VOCs have a high potential to produce low-volatility oxidation products (*47*) that will then easily condense and create or grow particulate matter (PM). Inhaling particulate matter, especially that with high organic content, is associated with a wide variety of respiratory and cardiovascular health effects (*48*). In addition, many of these VOCs, particularly the anthropogenic aromatic hydrocarbons, have high respiratory uptake and accumulate in human adipose tissue (*49*) which may lead to unfavorable health effects (*50*).

Example Exercises

Here are three examples of how air quality data can be used to look at issues of environmental justice.

1. California Air Resources Board (ARB) Ozone and NO_x Concentrations in Los Angeles

Before undertaking data analysis, students should prepare by engaging in background research. This is important not only to provide context for environmental issues, but also to demonstrate to students that there is a wealth of data freely available. This project is just an example of how one might use the information. Students can learn about the formation of tropospheric ozone, its seasonal trends, and related health concerns. They can also research the California ARB monitoring network and learn how the ozone instrumentation works. Please note that this exercise was created for use by students in Los Angeles, so it uses data from Southern California, but other air quality datasets can be used, as long as demographic data are available on a similar spatial scale.

In this example exercise, California ARB air quality data, specifically ambient ozone and NO_x concentrations, are extracted from specific measuring sites chosen from the AQMIS monitoring network (26). Ozone data are easily accessible; on the main AQMIS webpage, the "latest ozone" button reveals ozone data for the previous few days for each air quality region of California. Selecting "South Coast Air Basin" generates a table of annual ozone data for the chosen region. This table, and all subsequent tables, can be downloaded using one of the "Download Data" links at the bottom of the page. Each data point in the table is itself another link. By clicking on a specific year, a table of monitoring sites and maximum ozone concentrations for that year becomes available to download. Unfortunately, ozone data accessed this way are only available dating back to 1995. To obtain older data (dating as far back as ~1980, depending on the site), one must follow a procedure similar to that below for NO_x (using the "Special Reports" tab under the "Air Quality Data" category).

NO_x (amongst other pollutants, including ozone) data are found through the "Air Quality Data" button on the main AQMIS website. To obtain annual averages, one must use the "Special Reports" tab. Many parameters, including pollutant, year, and air basin, can be chosen here. To mirror the ozone data previously obtained, select "NO_x", "ppm", "South Coast" Basin, "Annual Statistics by Site", and sort by "Basin, County, Site"; this generates the specified data table. Unfortunately, at the present time there is apparently no way to download these data. One can use the toggle box at the bottom of the screen to change the year of displayed data, and then a multi-year dataset can be created (by hand).

Because we are using the data to explore environmental justice issues, the sites chosen must correspond with the demographic data to which it will be compared. This example project is based on data in Los Angeles, so we will use the simplified Census data from the Mapping L.A. Neighborhoods website (33). In the "Find Your Neighborhood" box on the webpage, "The complete list" opens the list of all 272 designated neighborhoods in Los Angeles County. Clicking on the name of each neighborhood reveals its characteristics. Any demographic variable could be added to the dataset (by hand). For this exercise, "median household income (in 2008 dollars)" is found under the "Income" heading.

After comparing the Mapping L.A. Neighborhood website with the AQMIS website, there are 8 L.A. neighborhoods that have long-term (since ~1980) ozone and NO_x data available. A spreadsheet of compiled data can be created by hand in Excel, divided by year, measurement site, and pollutant. Because some data are not easily downloaded, instructors may want to prepare a spreadsheet with the AQMIS data ahead of time, and then students can fill in the demographic information. If so, the assignment should still have students answer questions about the AQMIS dataset so that they understand its origin.

It is beneficial to get a sense of the general air quality trend. A plot of the ambient concentrations of the pollutants over the 30+ years clearly shows an improvement in Los Angeles' air quality (Figure 1); this is a consistent trend for each site. To look at how socioeconomic status plays in the air quality trend, the pollutant concentration is plotted against the median income of each neighborhood. Linear regression and correlation coefficients (R^2 and r values) for all 14 years are included in Table 1. Average results show a positive relationship between ozone concentrations and income. However, there is a very strong, statistically significant, negative relationship between NO_x concentrations and median income levels for most years. The income correlations with ozone are not as strong and this is likely due to air transport. As a secondary pollutant, there is more ozone in air downwind from NO_x sources. The variation in meteorology can then have a larger impact on secondary pollutant distribution than on that of primary pollutants. Trends in both NO_x and O_3 are significantly correlated with median 2008 income (Figure 2).

Table 1. Linear Regression and Correlation Coefficients between Ozone or NO_x and Median Household Income. Statistical Significance, $p<0.05$, Is Indicated by the Asterisks (*) (Critical Value = 0.707, Two-Tailed Test).

	Ozone			NO_x		
Year	R^2	r		R^2	R	
2000	0.087	0.29		0.43	-0.66	at p<0.10
2001	0.20	0.45		0.57	-0.75	*
2002	0.41	0.64		0.62	-0.79	*
2003	0.59	0.77	*	0.57	-0.76	*
2004	0.40	0.63		0.60	-0.77	*
2005	0.32	0.57		0.52	-0.72	*
2006	0.37	0.61		0.36	-0.60	
2007	0.036	0.19		0.58	-0.76	*
2008	0.55	0.74	*	0.54	-0.74	*
2009	0.15	0.39		0.53	-0.73	*
2010	0.42	0.65		0.35	-0.59	
2011	0.54	0.74	*	0.52	-0.72	*
2012	0.43	0.65		0.56	-0.75	*
2013	0.40	0.64		0.41	-0.64	
AVG	0.35	0.57		0.51	-0.72	*

The opposite trends for NO_x and O_3 in Los Angeles may surprise some students. This "teaching moment" can lead to a discussion that highlights the importance of understanding chemistry and transport in the study area. Because NO_x is the sum of both NO and NO_2, and because the primary pollutant, NO, quickly reacts to become NO_2, NO_x concentrations typically track combustion sources. On the other hand, as a secondary pollutant, O_3 is formed downwind from the combustion source. In Southern California, low income residents are generally exposed to significantly more primary pollutants, and higher income residents are generally exposed to more secondary pollutants; this conclusion was also reached by Marshall (51). The data suggest that low income neighborhoods are located near pollutant emission sources while more affluent neighborhoods are downwind from those sources. The correlation between exposure and income level reveals that not everyone is enjoying the same healthy environment.

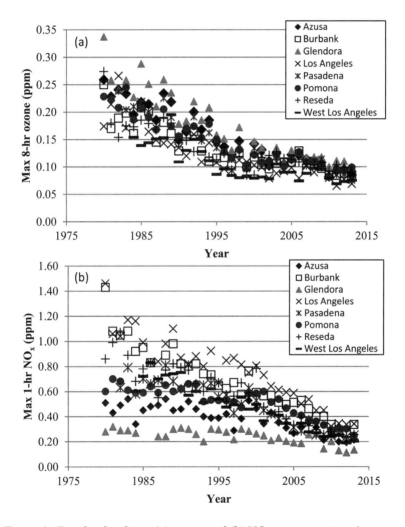

Figure 1. Trends of ambient (a) ozone and (b) NO_x concentrations (parts per million by volume, ppm) in eight Los Angeles neighborhoods from 1980 to 2013.

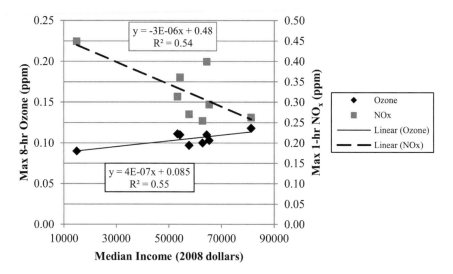

Figure 2. The relationship between ambient ozone and NO_x concentrations (parts per million by volume, ppm) and median income level in eight Los Angeles neighborhoods for 2008.

2. NASA Satellite Data and Google Earth

For a wider spatial distribution of data, NASA provides data for a broad range of air quality parameters that can be opened and layered in Google Earth (*52*). At the outset, students should learn about the data generally, including the instrumentation aboard the satellites. For example, if NO_2 was the pollutant of choice, students can learn about the Ozone Monitoring Instrument (OMI) (*53*) on the Aura satellite (*54*): i.e., how OMI uses spectroscopy to measure NO_2, what the physical size of the instrument is, how often Aura measures a particular area, etc.

In this example exercise, global carbon monoxide (CO) and nitrogen dioxide (NO_2) datasets are layered with population density. (The idea for layering NO_2 and population was inspired by Urban et al. (*55*)). All data files can be obtained through the NASA Earth Observations (NEO) website (*28*). "Carbon Monoxide" and "Nitrogen Dioxide" links are found under the "Atmosphere" tab while population density is found using the "Population" link under the "Life" tab. Each link will lead to another website from where the data can be downloaded. Scroll down the page to select the month and year of interest (January 2014 will be used here). For this exercise, select the "Google Earth" file type on the right side panel. Choose the highest resolution; for population density, the 0.1 degree resolution is selected by clicking on the "3600×1800" link. This will create a file with the ".kmz" extension that can be saved and opened in the Google Earth program. Higher resolution pollutant data with larger color gradients for better spatial pattern differentiation can be downloaded from websites specific to the instrumentation – Aura (*56*) for NO_2 and MOPITT (*57*) for CO. On the Aura validation data center website, simply select the month and year of interest (January 2014 will be used here) and click on the Google Earth symbol to download the Google Earth (.kml) file. On the MOPITT website, scroll down to find the "V6 TIR/NIR", "Monthly Plots" link. This link opens a new page where the month and year of interest can be selected with drop-down menus. Once the date is chosen, the dataset is displayed in graphic form; there is a link at the bottom of the page "MOPITT Monthly Google Earth File" where the (.kmz) file can be downloaded. Care must be taken to download pollution data files from the same time-frame for direct comparisons.

Once the files are opened in Google Earth, students can see pollution patterns. Geographical influences can be explored, as demonstrated by Urban et al. (*55*). Seasonal patterns can be elucidated with multiple months of data from a particular pollutant. Overall, it is quite striking how NO_2 and CO concentrations seem to visually correlate with population density – these pollutants are markers for combustion. It is more interesting to explore the anomalies. If students are directed to specific locations, they can be asked to explain the unusual pattern.

For CO, the most striking global anomaly in January is over sub-Saharan Africa. While the population density is relatively low, and the NO_2 concentrations are low as compared to industrialized cities, the carbon monoxide is much higher than what is observed in urban centers. What is unique about this location? It has many savannah fires. Whether naturally occurring or due to anthropogenic deforestation, fire is another significant combustion source; however, fire presents a different pollution emission profile than the typical internal combustion engine. The main concern with NO_2 pollution is the potential to form tropospheric ozone, while CO, on the other hand, has known direct health effects. Carbon monoxide interferes with a person's ability to deliver oxygen to the body's organs. The people of sub-Saharan Africa are generally economically disadvantaged; in fact all of the countries affected by the high CO levels rank as "low" on the United Nations Development Program's "Human Development Index" (*58*).

One nitrogen dioxide anomaly closer to home is in Morgan Lake, New Mexico. There is a lot of NO_2 present without a very large population density (Figure 3a). Students should likely predict that there would be a low concentration of NO_x, given the location and population density (Figure 3). Using the "Google Earth Community" and "Photos" layers, students can explore the region to try and find an explanation for the anomaly. Eventually, they will discover that the Four Corners Generating Station, a coal-fired power plant, is sitting at the edge of Morgan Lake. To highlight the environmental injustice, ask students who owns the land where the facility is built. A quick internet search reveals that the land is owned by the Navajo Nation. Historically, the Native American people have suffered great social and environmental injustices.

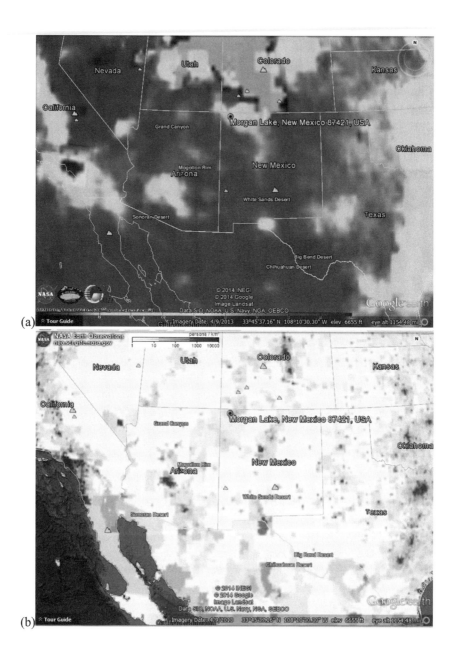

Figure 3. NASA data visualized in Google Earth. (a) Nitrogen dioxide (NO_2) from the Aura satellite and (b) population density distributions over the Southwest United States in January 2014. In both cases, the brighter the color, the higher the concentration. (see color insert)

Table 2. Linear Regression and Correlation Coefficients between Individual VOCs and the Percent of Households with Income Less than $20,000. VOCs Are Classified as (a) Anthropogenic or (b) Biogenic. Statistical Significance, p<0.05, Is Indicated by the Asterisks (*) (Critical Value = 0.755, Two-Tailed Test).

a) Anthropogenic	% <$20,000 income			temperature		humidity	
	R^2	r		R^2	r	R^2	r
4-ethyltoluene	0.79	0.89	*	0.0014	0.037	0.11	-0.33
1,3,5-TMB[a]	0.31	0.56		0.0079	0.089	0.37	-0.61
1,2,4-TMB[a]	0.55	0.74	p<0.10	0.076	-0.28	0.036	0.19
1,2,3-TMB[a]	0.92	0.96	*	0.026	-0.16	0.016	-0.13
1,2,4,5-TMB[b]	0.66	0.81	*	0.042	0.20	0.017	0.13
b) Biogenic	% <$20,000 income			temperature		humidity	
	R^2	r		R^2	r	R^2	r
borneol	0.0065	0.081		0.22	0.47	0.16	-0.40
α-pinene	0.024	0.16		0.024	0.15	0.10	0.32
verbenone	0.23	0.48		0.20	0.45	0.37	-0.61

[a] TMB: trimethylbenzene [b] TMB: tetramethylbenzene

3. Direct Gound VOC Measurements

Instead of relying on databases, a laboratory class can take their own air quality measurements. The example exercise presented here uses the HayeSep Q adsorbent method described by Bouvier-Brown et al. (*46*). Air was sampled for 3 hours using pre-made adsorbent cartridges (Analytical Research Systems, Inc.) and a portable pump/flow controller (SKC, Inc.) in public parks throughout the Los Angeles Basin. Care was taken so that samples were taken at the same time of day, and all within a two-week span, so that the measurements would not be significantly influenced by changes in meteorology. After sampling, the analytes were extracted and analyzed using a gas chromatograph-mass spectrometer (GC-MS). Peak areas were then calibrated using gas-phase standards (*46*). Because the project was in the Los Angeles Basin, demographic data was obtained from the Mapping L.A. database. VOC concentrations were plotted for each of the seven field sites and overlaid with the percent of households in that neighborhood whose income is less than $20,000 (Figure 4). The anthropogenic VOCs, all substituted benzenes likely directly emitted from fuel sources, showed a clear positive linear correlation with the percent of households in the neighborhood whose income is less than $20,000. The linear regression and correlation coefficients (R^2 and r values) are shown in Table 2. Three anthropogenic VOCs have a very strong, statistically significant positive relationship with the percent of low income residents in the neighborhood. Biogenic VOCs plotted with the income demographic data did not show a trend (Table 2, Figure 5); there is no significant relationship between the biogenic VOC concentrations and percent of households with low income. Linear regression coefficients were also obtained for additional independent variables, including temperature and humidity (Table 2). There is no significant correlation between the anthropogenic VOCs and temperature or humidity, but there is a weak correlation between these environmental parameters and the biogenic VOC concentrations. This makes sense because these factors control biogenic VOC emissions (*59*, *60*).

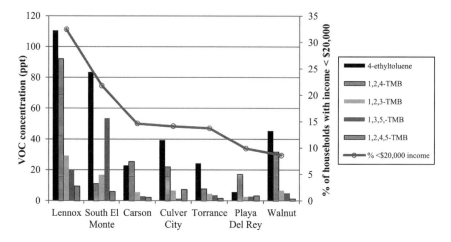

Figure 4. Anthropogenic VOCs quantified (parts per trillion by volume, ppt) at seven different sites in the Los Angeles Basin overlaid with the percent of households with less than $20,000 income. TMB: tri/tetramethylbenzene.

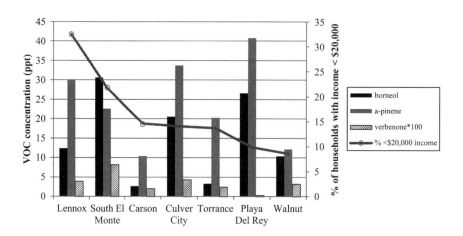

Figure 5. Biogenic VOCs quantified (parts per trillion by volume, ppt) at seven different sites in the Los Angeles Basin overlaid with the percent of households with less than $20,000 income. Verbenone concentrations are multiplied by 100 to fit on this scale.

Overall, there is a clear relationship between anthropogenic VOC concentrations and the income level of Los Angeles neighborhoods. There is a higher concentration of the directly emitted VOCs in low-income neighborhoods, and the trend is not apparent for the biogenic VOCs. Not all anthropogenic compounds behave the same, so it is important to have speciated data. The differences may have to do with sources or sinks of individual compounds, but more research is needed to "tease out" the specifics. Regardless, an experiment such as this not only develops students' analytical skills, but also provides a first-hand look at the disparate distribution of anthropogenic VOCs.

Summary

Air quality data, whether collected in the field or extracted from online databases, can be used to demonstrate exposure patterns. Evidence of a disparate exposure to a pollutant, correlated with socioeconomic disadvantage, brings attention to contemporary environmental justice problems. Educational research indicates that it is important for students to see the application of science to broader social issues. These exercises give students some tools to understand the impact on society that they can have through scientific study. Exercise 1 describes the use of data from Southern California, but many other air quality databases exist, of varying resolution, throughout the world. Further development of the three exercises could include an increased level of student control and inquiry; this would likely entice more civic action (*23*).

Acknowledgments

I would like to thank Erica Carrasco for collecting and analyzing the data collected around the Los Angeles Basin using the HayeSep Q cartridges. Erica was funded through the Loyola Marymount University (LMU) Rains Research Assistant Program. Work with the California Air Resources Board and NASA databases was spear-headed as a part of a course development grant funded by the LMU Seaver College of Science and Engineering. In addition, I acknowledge the support of the LMU Department of Chemistry and Biochemistry.

References

1. Cole, A. G. *J. Environ. Educ.* **2007**, *38*, 35–45.
2. Strife, S. *J. Environ. Educ.* **2010**, *41*, 179–191.
3. Skelton, R.; Miller, V. *The Environmental Justice Movement*; http://www.nrdc.org/ej/history/hej.asp (accessed Jul 16, 2014).
4. Clinton, W. J. *Federal Actions To Address Environmental Justice in Minority Populations and Low-Income Populations; Executive Order 12898* **1994**, *59*.
5. U.S. EPA *Environmental Justice*; http://www.epa.gov/compliance/ej/ (accessed Jul 16, 2014).
6. Brunekreef, B.; Holgate, S. T. *Lancet* **2002**, *360*, 1233–1242.

7. Hoek, G.; Krishnan, R. M.; Beelen, R.; Peters, A.; Ostro, B.; Brunekreef, B.; Kaufman, J. D. *Environ. Health* **2013**, *12*, 43.
8. Grineski, S. *Environ. Hazards* **2007**, *7*, 360–371.
9. Ritz, B.; Yu, F. *Environ. Health Perspect.* **1999**, *107*, 17–25.
10. Ritz, B.; Yu, F.; Chapa, G.; Fruin, S. *Epidemiology* **2000**, *11*, 502–511.
11. Wilhelm, M.; Ritz, B. *Environ. Health Perspect.* **2005**, *113*, 1212–1221.
12. Woodruff, T. J.; Parker, J. D.; Kyle, A. D.; Schoendorf, K. C. *Environ. Health Perspect.* **2003**, *111*, 942–946.
13. Morello-Frosch, R.; Pastor, M.; Sadd, J. *Ann. Am. Acad. Pol. Soc. Sci.* **2002**, *584*, 47–68.
14. Brainard, J. S.; Jones, A. P.; Bateman, I. J.; Lovett, A. A.; Fallon, P. J. *Environ. Plan. A* **2002**, *34*, 695–716.
15. Pearce, J.; Kingham, S. *Geoforum* **2008**, *39*, 980–993.
16. Miranda, M. L.; Edwards, S. E.; Keating, M. H.; Paul, C. J. *Int. J. Environ. Res. Public Health* **2011**, *8*, 1755–1771.
17. Yanosky, J. D.; Schwartz, J.; Suh, H. H. *J. Toxicol. Environ. Health. A* **2008**, *71*, 1593–1602.
18. Brochu, P. J.; Yanosky, J. D.; Paciorek, C. J.; Schwartz, J.; Chen, J. T.; Herrick, R. F.; Suh, H. H. *Am. J. Public Health* **2011**, *101*, S224–S230.
19. Morello-Frosch, R.; Pastor, M.; Sadd, J. *Urban Aff. Rev.* **2001**, *36*, 551–578.
20. National Research Council. *National Science Education Standards*; National Academy Press: Washington, DC, 1996.
21. National Research Council. *Inquiry and the National Science Education Standards*; National Academy Press: Washington, DC, 2000.
22. Markus, G. B.; Howard, J. P. F.; King, D. C. *Educ. Eval. Policy Anal.* **1993**, *15*, 410–419.
23. Morgan, W.; Streb, M. *Soc. Sci. Q.* **2001**, *82*, 154–169.
24. Warren, K. *J. Exp. Educ.* **1996**, *19*, 135–140.
25. California Air Resouces Board. *California Air Quality Data Products That Can Be Ordered*;http://www.arb.ca.gov/aqd/order/prodinfo.htm (accessed Jul 16, 2014).
26. *Air Quality and Meterological Information System (AQMIS)*; http://www.arb.ca.gov/aqmis2/aqmis2.php (accessed Sep 25, 2014).
27. *iADAM: Air Quality Data Statistics*; www.arb.ca.gov/adam (accessed Sep 25, 2014).
28. *NASA Earth Observations*; http://neo.sci.gsfc.nasa.gov/ (accessed Jul 16, 2014).
29. *Giovanni-The Bridge Between Data and Science*; http://disc.sci.gsfc.nasa.gov/giovanni (accessed Sep 25, 2014).
30. *Socioeconomic Data and Applications Center (SEDAC)*; http://sedac.ciesin.columbia.edu/ (accessed Jul 16, 2014).
31. Hajat, A.; Diez-Roux, A. V; Adar, S. D.; Auchincloss, A. H.; Lovasi, G. S.; O'Neill, M. S.; Sheppard, L.; Kaufman, J. D. *Environ. Health Perspect.* **2013**, *121*, 1325–1333.
32. Pastor, M.; Morello-Frosch, R.; Sadd, J. L. *J. Urban Aff.* **2005**, *27*, 127–148.
33. Los Angeles Times. *Mapping LA Neighborhoods*; http://maps.latimes.com/neighborhoods/ (accessed Sep 25, 2014).

34. Rubin, M. B. *Bull. Hist. Chem.* **2002**, *27*, 81–106.
35. Targa, J.; Loader, A. *Diffusion Tubes for Ambient NO₂ Monitoring: Practical Guidance for Laboratories and Users*; AEAT/ENVR/R/2504, AEA Energy & Environment, 2008.
36. U.S. EPA. *DRAFT Roadmap for Next Generation Air Monitoring*; 2013.
37. Piedrahita, R.; Xiang, Y.; Masson, N.; Ortega, J.; Collier, A.; Jiang, Y.; Li, K. *Atmos. Meas. Technol. Discuss.* **2014**, *7*, 2425–2457.
38. Snyder, E. G.; Watkins, T. H.; Solomon, P. A.; Thoma, E. D.; Williams, R. W.; Hagler, G. S. W.; Shelow, D.; Hindin, D. A.; Kilaru, V. J.; Preuss, P. W. *Environ. Sci. Technol.* **2013**, *47*, 11369–11377.
39. Król, S.; Zabiegała, B.; Namieśnik, J. *Trends Anal. Chem.* **2010**, *29*, 1101–1112.
40. Kumar, A.; Víden, I. *Environ. Monit. Assess.* **2007**, *131*, 301–321.
41. Camel, V.; Caude, M. *J. Chromatogr., A* **1995**, *710*, 3–19.
42. U.S. EPA. *Compendium of Methods for the Determination of Toxic Organic Compounds in Ambient Air*, 2nd ed.; Compendium Method TO-17 Determination of Volatile Organic Compounds in Ambient Air Using Active Sampling Onto Sorbent Tubes, EPA/625/R-96/010b, Cincinnati, OH, 1999.
43. National Institute for Occupational Safety and Health. Method 1501, Issue 3, *NIOSH Manual of Analytical Methods (NMAM)*, 4th ed., 2003 *127*.
44. Ramírez, N.; Cuadras, A.; Rovira, E.; Borrull, F.; Marcé, R. M. *Talanta* **2010**, *82*, 719–727.
45. Ras, M. R.; Borrull, F.; Marcé, R. M. *Trends Anal. Chem.* **2009**, *28*, 347–361.
46. Bouvier-Brown, N. C.; Carrasco, E.; Karz, J.; Chang, K.; Nguyen, T.; Ruiz, D.; Okonta, V.; Gilman, J. B.; Kuster, W. C.; DeGouw, J. A. *Atmos. Environ.* **2014**, *94*, 126–133.
47. Donahue, N. M.; Robinson, A. L.; Stanier, C. O.; Pandis, S. N. *Environ. Sci. Technol.* **2006**, *40*, 2635–2643.
48. Russell, A. G.; Brunekreef, B. *Environ. Sci. Technol.* **2009**, *43*, 4620–4625.
49. Järnberg, J.; Johanson, G.; Löf, A. *Toxicol. Appl. Pharmacol.* **1996**, *140*, 281–288.
50. Gerarde, H. *AMA Arch. Ind. Heal.* **1959**, *19*, 403–418.
51. Marshall, J. D. *Atmos. Environ.* **2008**, *42*, 5499–5503.
52. *Google Earth*; https://www.google.com/earth/ (accessed Jul 16, 2014).
53. *Ozone Monitoring Instrument (OMI)*; http://aura.gsfc.nasa.gov/scinst/omi.html (accessed Sep 25, 2014).
54. *Aura*; http://aura.gsfc.nasa.gov/scinst/index.html (accessed Sep 25, 2014).
55. Urban, M. J.; Bojkov, B.; Carter, B.; Dogancay, D.; Fermann, E. In *Earth Exploration Toolbook;* Science Education Resources Center at Carleton College, 2008; http://serc.carleton.edu/eet/aura/index.html (accessed Jul 16, 2014).
56. NASA Goddard Flight Center. *Aura Validation Data Center*; http://avdc.gsfc.nasa.gov/?site=705441739 (accessed Jul 16, 2014).
57. NCAR Atmospheric Chemistry Division. *Measurements of Pollution in the Troposphere (MOPITT)*; https://www2.acd.ucar.edu/mopitt (accessed Sept 25, 2014).

58. United Nations Development Programme. *Human Development Report 2013*; New York, 2013.
59. Guenther, A. B.; Zimmerman, P. R.; Harley, P. C.; Monson, R. K.; Fall, R. *J. Geophys. Res.* **1993**, *98*, 12609–12617.
60. Schade, G.; Goldstein, A.; Lamanna, M. *Geophys. Res. Lett.* **1999**, *26*, 2187–2190.

Chapter 8

Using Service Learning To Teach Students the Importance of Societal Implications of Nanotechnology

A-M. L. Nickel* and J. K. Farrell

Milwaukee School of Engineering, 1025 N Broadway, Milwaukee, Wisconsin 53202
*E-mail: nickel@msoe.edu.

As the field of nanotechnology develops, resources have been dedicated to studying the societal implications of nanotechnology therefore it is an important topic for an undergraduate course in nanotechnology. A service-learning course project provides students with an opportunity to engage in informing the public because students present topics of nanotechnology to small groups at a local high school. Prior to their service-learning project, students have built a solid understanding of the field and have participated in other assignments that established a theme of the societal implications of nanotechnology. The service-learning course project provides a meaningful learning opportunity that reinforces both the core concepts and the societal implications of nanotechnology.

Introduction

Nanotechnology is a growing field with great promise in coming years and decades. The field has impacted nearly every sector of industry and technology including cosmetics, clothing, information technology, food safety, medicine and energy (*1, 2*). These industries have developed more than 800 commercially available products in the last couple of decades (*3*). It has revolutionized research due to its interdisciplinary nature of bringing together scientists and engineers from a broad range of disciplines. Its breadth and depth make it an important topic to include in undergraduate science and engineering programs.

© 2014 American Chemical Society

Nanotechnology involves controlling, studying and employing materials that measure between 1 and 100 nm in at least one dimension. At this size scale, new physical, chemical, mechanical, optical and biological properties emerge that make materials different or better than the same composition at the microscale or larger scales (*4*). Because of the field's large impact, researchers have addressed the need for the societal impacts of nanotechnology to be studied as well. The National Nanotechnology Initiative (NNI) is a collective association involving the U.S. government agencies with nanotechnological research and development components (*5*). The National Science Foundation (NSF), the National Institutes of Health (NIH), the Environmental Protection Agency (EPA), the Department of Education (ED), and the Food and Drug Administration (FDA) are some of the agencies involved in the NNI. One of the goals of the NNI demonstrates the importance of the societal impacts of nanotechnology, "Support responsible development of nanotechnology" and the outcomes for this goal involve sustainability and evaluating risks and benefits of nanotechnology on society (*6*). Member agencies of the NNI have allocated funds for studying societal impacts such as the NSF's foundation of several large centers at universities across the country. Societal impacts of nanotechnology are recognized as important to study and are supported financially by U. S. government agencies. A report sponsored by the NNI in 2010 stated, "A key issue for academia, industry and government is to effectively communicate, inform and involve public participation in the dialogue on beneficial implications of nanotechnology the potential for risk and what is being done to ensure safe implementation of the technology" (*4*). Just as the NNI recognizes the significance of societal impacts of nanotechnology, so should a course dedicated to teaching students about nanotechnology.

This chapter aims to illustrate how a service-learning student project is the ideal assignment to bring significant meaning to students' understanding of the importance of the societal implications of nanotechnology. Service learning has been defined as a "course-based, credit-bearing educational experience in which students (a) participate in an organized service activity that meets identified community needs and (b) reflect on the service activity in such a way as to gain further understanding of course content, a broader appreciation of the discipline, and an enhanced sense of civic responsibility (*7*)." Specific goals and benefits of service learning to the nanotechnology students include the opportunity to:

- Repurpose their knowledge of nanotechnology to help a nonexpert seeking information;
- Gain greater understanding of the course material after teaching others;
- Engage in societal impacts of nanotechnology by educating the non-experts;
- Serve as role-models as college students, engineering students, and successful peers

A service-learning project is the ideal method to allow students to gain a meaningful understanding of the societal implications of nanotechnology. The service-learning project that will be discussed in this chapter allowed the nanotechnology students to educate non-experts in society about nanotechnology which is one of the challenges that nanotechnology experts identify as a need in addressing societal implications of nanotechnology.

Service Learning To Enhance Societal Impacts of Nanotechnology

Susanna Priest, an expert on public perception of nanotechnology describes three elements that need to be evaluated in assessment of the societal impacts of nanotechnology: they are risk analysis, risk perception and risk communication (*8*). Risk analysis attempts to quantify the harm that might result. Risk perception analyzes what society perceives as the harm. Once risk analysis and risk perception have been determined, risk communication is the communication between the communities of the developing technology and the general public. Education of non-experts is a component of this communication, however also important is having the experts hear the concerns and opinions of the non-experts. Thus, Priest emphasizes that communication must be a dialogue of listening and learning from both sides. Students should recognize that they might be the experts one day and that society might look to them for answers. They may also represent those who need to be asking the critical questions of the scientists. Both roles are valuable for the development of a field to have beneficial impacts on society while minimizing damaging effects. A service-learning assignment that allows students to educate others in the community about nanotechnology, provides them the opportunity to be experts and practice communicating complicated science to the public. John Dewey, one of the premier educational reformers of the 20th century, emphasized meaningful activity and stressed that communication in personal discourse is critical to learning from experiences and that "The values, aim, and expected response of others play a critical role in stimulating revised interest in each participant (*9*)."

Pedagogy Supporting the Incorporation of Service Learning

The publication of Ernest Boyer's 1990 special report *Scholarship Reconsidered: Priorities of the Professoriate* is a watershed moment in higher education. In this treatise, Boyer challenges the conventional idea that research and pedagogy are necessarily discrete entities and instead lays out four "separate, yet overlapping functions" that make up the work of the professoriate. Those four functions are well-known and have been cited a myriad of times since 1990, but it is worth listing them one more time since they form the framework from which many pedagogical experiments have sprung. The four functions are scholarship of discovery, scholarship of integration, scholarship of application, and scholarship of teaching (*10*). By redefining how educators may approach their scholarship and their classrooms, Boyer opened the door for interdisciplinary pedagogies, blended assignments, and actual or experiential learning. Just a few years after Boyer's publication, the Kellogg Commission formed in order to address what they deemed to be necessary changes to higher education.

The Kellogg Commission agreed with Boyer and devised seven characteristics of what makes for effective societal engagement on the part of universities: responsiveness, respect for partners, academic neutrality, accessibility, integration, coordination, and resource partnerships. From 1996 to 2001 the Kellogg Commission in conjunction with the National Association of State Universities and Land-Grant Colleges held meetings on the campuses of 25 land-grant institutions throughout the United States. The goal of these meetings was to assess and to implore higher education to "return to its roots" (*11*). The goal of the Kellogg Commission was to revitalize and revolutionize higher education by bringing it back to its origin point as integral to the larger community. The Kellogg Commission feels that "The engaged institution must:

- Be organized and respond to the needs of today's students and tomorrow's;
- Bring research and engagement into the curriculum and offer practical opportunities for students to prepare for the world they will enter;
- Put its resources - knowledge and expertise - to work on problems that face the communities it serves (*12*)."

In 2006, the Kellogg Commission revisited its decade old crusade and learned from the self-reporting of the 25 involved institutions. Included in the list of action commitments cited in that report were: "address the academic and personal development of students in a holistic way" and "strengthen the link between discovery and learning by providing more opportunities for hands-on learning, including undergraduate research (*12*)." Holistic learning focuses on education inside and outside of traditional educational settings (labs and classrooms) that engages student development socially, intellectually and academically (*12*). Often holistic learning takes the form of service-learning assignments, but it also creates a space in higher education for interdisciplinary courses and assignments that serve to engage the students, to inform the scholarship of educators, and to connect to the community.

Reflecting the ideas of the Kellogg Commission is *The Centrality of Engagement in Higher Education* by Hiram E. Fitzgerald *et al*. The core argument of this article is that public institutions have to move more toward a community relationship, especially as the role of the university changes (*11*). When Boyer redefined scholarship by suggesting it should include discovery, integration, application, and teaching so that what happened in the classroom would be treated as equal to traditional research. "Boyer challenged higher education to renew its covenant with society and to embrace the problems of society in shared partnerships with communities (*13*)." Interdisciplinary teaching and service learning are two key components of this community-engaged university.

Incorporation of Societal Impacts of Nanotechnology into a Course

Service learning meets Boyer's challenge to educators to embrace problems of society in partnerships. In the course entitled, Nanoscience and Nanotechnology the college students address a problem of the public which is lack of understanding of nanotechnology by reaching out to educate high school students. In doing this, students are participating in educating the nonexperts which Priest identifies as an important component to the societal implications of nanotechnology (*8*). In addition, they are educating students who might be motivated to learn more about nanotechnology or someday participate in a career that advances the field.

Societal impacts of nanotechnology are included among the topics of this course. The course covers an introduction to the science and technology of the size scale between 1 and 100 nm and, besides societal impacts, also covers self-assembly; comparison of properties that change from the bulk to the nanoscale; allotropes of carbon such as graphene, carbon nanotubes and buckyballs; nanoparticles; tools of nanotechnology; and applications and devices incorporating nanotechnology. These topics lay a foundation for the major topics in nanotechnology and provide examples of the future potential of the emerging field. The topic of societal impacts of nanotechnology grows as an underlying theme to the course. It is developed primarily through reading assignments but culminates in a service-learning project and is summarized in classroom discussions that complete the course.

The course has five components related to societal impacts. First, students gain an understanding of nanotechnology and its potential growth based on its diversity and exciting new material properties. Second, students experience a new technology and its impact on society using an example from science fiction literature as a case study. Third, students are exposed to the idea of environmental impacts of nanotechnological waste. Fourth, students engage with public perception of nanotechnology by playing a game called NanoVenture. Lastly, students practice communication with the public about nanotechnology where they serve as experts and they educate nonexperts about an aspect of nanotechnology. The course concludes with a discussion centered on the societal impacts of nanotechnology and the students' experiences in their service-learning projects conclude the course.

The first introduction to societal impacts of nanotechnology occurs during the first or second week when students read and discuss the *Point/Counterpoint: Nanotechnology* article from *Chemical and Engineering News* with Richard E. Smalley and K. Eric Drexler discussing their disagreements about building nanomaterials (*14*). The article provides an opportunity to discuss a brief history of nanotechnology including the contributions of the authors to the field and to a scientist to whom they refer, Richard Feynman, who is often credited as the grandfather of nanotechnology. In this article, both Smalley and Drexler are primarily engaged in debating the science behind mass producing nanomaterials from the bottom-up. However they also mention the importance of messaging to the public and the harm that inappropriate messaging can have. The article provides students with an introduction to the idea that science fiction and the media can impact the public's perception of nanotechnology. Scientists and engineers care about the public's perception of their fields because unfavorable perception can hinder progress and development. Additional exercises have included showing students pictures of "nanobots" or "nanites" from pop culture sources such as television or movies and asking students to project what Smalley or Drexler would have thought of the incorporation of them into pop culture.

Just as scientists ask "What if a material could?", science fiction authors write "What if a material did?" Because both members ask related questions, science fiction can be used to allow students to see the importance of Priest's three elements of the societal impacts of nanotechnology – risk analysis, risk perception and risk communication. The literature aids the students in evaluating the risk analysis conducted, identifying what the experts know and what the nonexperts should know, identifying the concerns of the nonexperts and pin-pointing topics of discussion that should occur. In a 2007 chapter published by Educause titled *Authentic Learning* the emphasis is on learning that ". . .focuses on real-world, complex problems and their solutions, using role-playing exercises, problem-based activities, case studies, and participation in virtual communities of practice (*15*)." One aspect that is often overlooked or underdeveloped in interdisciplinary courses is reading comprehension (*16*). Students need to be taught to read with an open mind so that underlying assumptions and main messages can be better understood. Including instructions alongside reading material can help students become better readers. A second way to make students become better readers is through role play wherein students learn the difference between critiquing a text and understanding a text (*16*). Students are often quick to criticize a work without understanding the text's true goals and underlying assumptions. By not truly reading and understanding a text, students often miss important information that will inform assignments down the road. An activity related to role play involves students predicting what Smalley and Drexler might think about a new material or device. In order to truly engage students, this course employs interdisciplinary assignments in order to get students to think about nanotechnology in terms of ethics, the environment, the future, as well as through the lens of scientific inquiry.

As part of the interdisciplinary approach of the nanotech course, a science fiction scholar is invited in and presents an introduction to the function and role of science fiction literature in scientific and cultural discussions. Students are better informed to read science fiction in context after learning the basics of science fiction. Students are assigned to read *Prey* by Michael Crichton. The novel has many attributes that work well for this class. Crichton novels are typically engaging, accessible and popular among engineering students. Even the students who do not commonly read novels for pleasure enjoy reading the book. More complex and deep examples of science fiction exist, but the level is appropriate for this course because the students are still expected to read textbook content, complete homework assignments and study the basics of nanotechnology. *Prey* serves two roles in the course. The first role is to allow students to demonstrate what they have learned about nanotechnology by evaluating Michael Crichton's portrayal of nanotechnology. The second role is to bring up aspects of the societal impacts of nanotechnology.

As a science fiction author, Crichton researched nanotechnology and then wrote a novel where the nanotechnology is not adequately controlled or contained and has a detrimental impact on society as it runs amok. When students read the novel, they can evaluate Crichton's portrayal of nanotechnology in two ways. First they can evaluate the accuracy of the nanotechnology – where is the actual field of nanotechnology in its development compared to the nanotechnology described in the book and what are the possible mistakes in the science. Secondly students encounter societal implications of nanotechnology. Examples include the conversion of a noble pursuit into a product to make money; a research team that gets so involved with the development of the research that they stop questioning safety; environmental impacts and individuals' health compromised. Experiencing the societal implications of nanotechnology in a fictional world presents students with an authentic learning opportunity in a relatively safe environment. Students can discuss mistakes and make accusations of fictional characters without threatening or challenging real people or events. Since *Prey* is hard science fiction, Crichton is able to create a real enough scenario to be treated as a case study and therefore using it in the classroom impacts learning. Students can practice risk analysis, and they can consider ways to maximize beneficial impacts and minimize the detrimental impacts on society.

Midway through the course, students are assigned to write an essay that evaluates how Crichton portrayed nanotechnology which gives students an opportunity to demonstrate what they have learned about nanotechnology. For example students have criticized Crichton because at one point he gives dimensions of the nanobots that aren't possible relative to the size of atoms. Students have also praised Crichton for using an example of bottom-up approach to building the nanobots. Many students recognize that there are cases where the nanotechnology strays from reality in order for Crichton to build his plot and others recognize that his portrayal of nanotechnology could cause fear or distrust in the field; both topics have fueled classroom discussions.

Reading and evaluating *Prey* provides students with a fictional example of nanotechnology and its impact on society. The novel uses examples that they have witnessed in pop culture such as "nanobots" or "nanites." There are clear examples of the areas of societal implications of nanotechnology including ethical, legal and environmental implications as well as public perception of nanotechnology. Classroom discussions or student assignments have involved students finding examples of where society was impacted by nanotechnology from the novel. Students find examples that fit under the categories of ethical, legal, environmental implications as well as public perception of nanotechnology such as, "What is the impact on the environment?" and "How can a company allow nanobots to escape?" Students could be further challenged to find places in the novel where risk analysis, risk perception and communication were conducted or ignored.

A subtopic of societal impacts of nanotechnology, that of environmental impacts, is highlighted. Articles related to risk analysis of nanotechnology have been incorporated into classroom discussions and assignments such as articles on the impact of silver nanoparticles on plant growth or medical treatment of cancer using gold nanoparticles (*17*, *18*). In the case of the silver nanoparticles article, students read the article and ask questions of the authors. One of the authors of the paper is a colleague and so she discusses answers to their questions as a guest lecturer.

Another opportunity for students to explore societal implications of nanotechnology occurred when students played the board game, NanoVenture, in class. In this game, students are running a country and they buy, rent and sell nanotechnologies. The decisions that they make during the game have financial costs and benefits, but those decisions also have gains and losses in public approval ratings as well. Students can lose the game financially or by loss of public support. In contrast, players can earn more money with high approval ratings. The game educates students about specific nanotechnologies and describes the potential impact of decisions on public perception of nanotechnology *(19)*.

Midway through the course, students are aware and considering the societal impacts of nanotechnology as they continue to learn about the field. These examples don't allow students to engage in affecting public. At this point, their service-learning course project starts holding greater meaning. They are called to engage with society to teach others about nanotechnology, thus educating non-experts as Priest described *(8)*. This element provides a perfect opportunity for students to get involved in societal impacts with a service-learning assignment to educate the public about nanotechnology. Without an understanding, the public cannot ask the appropriate ethical, legal and environmental questions that will impact lives of its individuals.

Service Learning To Strengthen the Theme of Societal Impacts of Nanotechnology

Service learning allows students to serve as experts and begin the communication between experts and non-experts as they take their part in educating society. Service learning enhanced students' understanding of the material by requiring them to engage in the material as other active learning strategies do. The service-learning assignment plays two roles in this course. The first is to solidify students' understanding of specific aspects of nanotechnology following the idea that in order to educate others, one must have a broad and deep understanding first. Students conduct a thorough investigation of a topic in order to develop an effective strategy for teaching it. The second role of service learning in this course is to engage in the communication of nanotechnology to others. Far too many members of the public do not have even a general definition of nanotechnology. In effort to engage society in conversations about the impacts of nanotechnology, society must be informed. Communication requires educating the nonexperts. When society is informed it is much better positioned to engage in conversations about societal impacts of nanotechnology. Furthermore, if the field is to develop, future scientists and engineers are needed to study, develop and employ new nanotechnological materials and products. Students have potential for engaging other students about nanotechnology because they speak to each other as peers and even role models.

Service Learning Assignment and Course Details

The course is comprised of mostly upper level undergraduate engineering students from particularly architectural, biomedical, electrical, industrial, and mechanical engineering programs. The course is often chosen to fulfill a science-elective for these students and is categorized as a basic science course using the *Accreditation Board for Engineering and Technology (ABET)* definitions. Some students enroll in the course to earn credits toward one of the science minors offered. One of the significant challenges to teaching the course is the broad range of backgrounds in science that the students have and recall. The second is that the Milwaukee School of Engineering (MSOE) is on a trimester system which means there are 10 weeks in a term as opposed to the standard 15 that schools on semester systems have. This means that the service-learning project has a very compact time frame and demands a great deal of focus, concentration, and commitment on the part of the student. A third and final trial is that it can be challenging to engage students since they might prioritize finding jobs or internships, completing capstone projects, or focusing on courses in their program higher than involvement in their science elective on nanotechnology. The service-learning activities can have greater impact on learning and this course's service-learning project builds on that foundational pedagogy.

Kolb and Kolb performed a multi-faceted study that examined the ways in which varied learning styles and environments could be combined to provide the best experiential learning in an educational setting. They reported their results in the article *Learning Styles and Learning Spaces: Enhancing Experiential Learning in Higher Education.* The Kolbs argue that the entirety of a learner's physical and social environment must be taken into consideration when venturing out into an experiential learning exercise. At the base of the learning is respecting what the student already knows about the subject matter. "Instead the effective teacher builds on exploration of what students already know and believe, on the sense they have made of their previous concrete experiences (*20*)." The social interactions that the learner will engage in are vital to experiential learning and also serve to define a learner's experience. Group work can often cause personal biases and latent hostilities to rise to the surface if the instructor is not adequately engaged with the learning environment. "To learn requires facing and embracing differences; whether they be differences between skilled expert performance and one's novice status, differences between deeply held ideas and beliefs and new ideas, or differences in the life experience and values of others that can lead to understanding them (*20*)." For Kolb and Kolb experiential learning is not complete until the learner engages in a thoughtful reflection of the entire exercise from receiving the assignment to turning in the final product.

Students have found their service-learning course project to be engaging even when they do not show motivation for the course material. The course project incorporates service learning and the idea that it is important for the public to have some understanding of the new technologies that emerge such as nanotechnology so that society can have informed discussions about societal implications of nanotechnology. In groups of five to six students, students are asked to develop a ten minute presentation with at least one hands-on, table-top demonstration that will help teach the general public about some aspect of nanotechnology. Their presentations aim to be engaging and informative with interactive components. In developing their presentations, students make efforts to teach the basic concepts needed to understand their demonstrations and activities. Some groups have chosen to have activities that show the size scale. Others built large demonstrations of scanning tunneling microscopes or self-assembly. Groups have included games, demonstrations, activities, slides, animations and movies into their presentations to students.

The audience for these presentations has been science students of St. Joan Antida High School (SJA). SJA is an urban all-girls high school located in downtown Milwaukee within walking distance to campus that serves both a culturally and economically diverse student population. The ethnicity demographics for SJA are 58% African American, 33% Latina, 5% Caucasian, 3% Asian and 1% American Indian. Of the 250 female students, 98% receive Milwaukee Choice tuition dollars and 98% receive free- or reduced-priced lunch (*21*). SJA has a strong program in science, technology, engineering and math and also involves a significant number of students in Project Lead the Way courses. The population of SJA represents underrepresented groups in science and engineering and therefore is potentially valuable group to get excited about nanotechnology. In contrast, MSOE students are 23% female, 13% minorities and 8% international students (*22*).

Progress reports are submitted prior to the presentation for the instructor to gauge progress and to identify issues. Topics have not needed to be assigned to students because enough variation occurred naturally. Time constraints prevent every SJA student from hearing every nanotechnology presentation so some repetition can be avoided.

Reflection papers are written by the MSOE students following their presentations (*7*). The reflection provides an opportunity for students to revisit their experience and evaluate its impact on the SJA students and themselves. In these reflections, students might describe frustrations with the logistics or the group work, but most praise the presentations to SJA as a valuable experience.

Class schedules at both institutions limit the duration and frequency of interaction that can occur between the MSOE and SJA students. Group size is also dictated by the logistical issues such as time and space for presentations. The SJA students enjoy the presentations and are engaged enough to ask questions of the MSOE students. The SJA students' questions are about nanotechnology and the content in the presentations as well as questions about college life, engineering, and MSOE. Each year SJA students were able to interact with 3-6, ten-minute presentations.

The course instructor and the high school teacher find this activity to be a worthwhile exercise and worthy of the time used to implement the assignment. Specifically, SJA students are provided the opportunity to:

- Interact with college engineering role models;
- Witness enthusiasm for nanotechnology from peers;
- Learn about nanotechnology, engineering and college life from peers;

The authors have yet to measure the service-learning project's impact on learning; however, the authors have witnessed students synthesizing and analyzing class material and disseminating that content to their peers. This evidence for student development has motivated the instructors to continue including the project in the course. In addition, the instructors witness engagement of both the SJA students and the nanotechnology students during presentations. Student reflection papers also recognize the value of the assignment. Future work will include measurement of the impact and learning.

As a culminating project for the course, this service-learning project provides students with the opportunity to share their knowledge with others and it provides students the opportunity to engage in societal impacts of nanotechnology. Further content on societal impacts close the course, therefore, the course ends with a topic that has a significant impact on the students as members of society.

Conclusion

Allowing students the opportunity to teach other students about nanotechnology engages them in the content of the course. They investigate and explore an area of nanotechnology to reinforce the basic concepts of nanotechnology. Using their understanding of the course content and the underlying theme of societal implications of nanotechnology as a foundation, they are prepared to serve as experts on nanotechnology. As they engage in educating others, they also participate in societal impacts of nanotechnology because they serve as experts and educate the nonexperts. Students leave the course with meaningful understanding and holistic learning of the concept, that in order for society to ask the right questions of the developing technology, society must be informed. This course and specifically the service-learning assignment prepare students to extend this concept beyond nanotechnology to any new developing technology that they might encounter in their personal or professional lives.

References

1. Hornyak, G. L.; Dutta, J.; Tibbals, H. F.; Rao, A. K. *Introduction to Nanoscience.*; CRC Press: New York, 2008; pp 50−52.
2. Cademartiri, L.; Ozin, G. A. *Concepts of Nanochemistry*; Wiley-VCH: Weinheim, 2009; pp 1−4.
3. *Nanotechnology Products, Applications & Instruments* (n.d.). Nanowerk; accessed June 19, 2014, http://www.nanowerk.com/nanotechnology/ nanomaterial/products_a.php.
4. Roco, M. C.; Mirkin, C. A.; Hersam, M. C. *Nanotechnology Research Directions for Societal Needs in 2020*; Springer: Boston, MA, 2010.
5. *National Nanotechnology Initiative Home Page* (n.d.); accessed June 10, 2014, National Nanotechnology Initiative: http://www.nano.gov.
6. *NNI Vision, Goals and Objectives* (n.d.); accessed June 10, 2014 National Nanotechnology Initiative: http://www.nano.gov/about-nni/what/vision-goals.

7. Bringle, R. C.; Hatcher, J. A. Reflections in Service-Learning: Making Meaning of Experience. In *Introduction to Service-Learning Toolkit: Readings and Resources for Faculty*; Campus Compact: Boston, MA, 2003; pp 83–90.
8. Priest, S. H. *Nanotechnology and the Public: Risk Perception and Risk Communication (Perspectives in Nanotechnology)*; CRC Press: Boca Raton, FL, 2011; pp 3–36.
9. *The Moral Writings of John Dewey*; Gouinlock, J., Ed.; Prometheus Books: New York, 1994; xxxvi.
10. Boyer, E. L. Scholarship Reconsidered: Priorities of the Professoriate. *The Carnegie Foundation for the Advancement of Teaching*; Wiley Jossey-Bass: New York, 1990.
11. Kellogg Commission on the Future of State and Land-Grant Universities. *RETURNING TO OUR ROOTS: Executive Summaries of the Reports of the Kellogg Commission on the Future of State and Land-Grant Universities*; 2001; accessed May 30, 2014, http://www.aplu.org/page.aspx?pid=305.
12. Byrne, J. V. *Public Higher Education Reform Five Years After The Kellogg Commission on the Future of State and Land-Grant Universities*; 2006; accessed May 30, 2014, http://www.aplu.org/page.aspx?pid=305.
13. Fitzgerald, H. E.; Bruns, K.; Sonka, S. T.; Furco, A.; Swanson, L. The Centrality of Engagement in Higher Education. *J. Higher Educ. Outreach Engagement* **2012**, *16*, 7–27.
14. Baum, R. M. Point-Counterpoint: Nanotechnology. *Chem. Eng. News.* **2003**, *81* (48), 37–42.
15. Lombardi, M. M. Authentic Learning for the 21st Century: An Overview. *Educause*; Oblinger, D. G., Ed.; 2007.
16. Andersson, A.; Kalman, H. Reflections on Learning in Interdisciplinary Settings. *Int. J. Teach. Learn. Higher Educ.* **2010**, *22*, 204–208.
17. Wang, J.; Koo, Y.; Alexander, A.; Yang, Y.; Westerhof, S.; Qingbo, Z. Phytostimulation of Poplars and Arabidopsis Exposed to Silver Nanoparticles and Ag+ at Sublethal Concentrations. *Environ. Sci. Technol.* **2013**, *47*, 5442–5449.
18. Jacoby, M. The Mystery of Hot Gold Nanoparticles. *Chem. Eng. News.* **2013**, *91* (13), 44–45.
19. *Hands On Science Kits and Demos*, (n.d.); accessed June 10, 2014, from Institute for Chemical Education, http://ice.chem.wisc.edu/Catalog/SciKits.html.
20. Kolb, A. Y.; Kolb, D. A. Learning Styles and Learning Spaces: Enhancing Experiential Learning in Higher Education. *Acad. Manage. Learn. Ed.* **2005**, *4*, 193–212.
21. *Quick Facts About St. Joan Antida High School*, (n.d.); accessed May 29, 2014, St. Joan Antida High School: http://saintjoanantida.org/index.php/en/.
22. *Who We Are*, (n.d.); The Milwaukee School of Engineering; accessed June 20, 2014, http://www.msoe.edu/community/about-msoe/who-we-are/page/1269/about-us-who-we-are.

Chapter 9

XRF Soil Screening at the Alameda Beltway

Steven Jon Bachofer*

Department of Chemistry, Saint Mary's College of California,
Moraga, CA 94575
*E-mail: bachofer@stmarys-ca.edu.

The 2012 Environmental Chemistry class at Saint Mary's College of California had the opportunity to screen the soils of a proposed community garden site for the City of Alameda. In collaboration with a community partner organization, the Alameda Point Collaborative, a mini-research project was integrated into a two-week environmental chemistry lab. Students had a two week instructional lab early in the course to learn how to use the instrumentation and to prepare a field site for sampling where the pollution source was well defined, so four weeks in total were devoted to XRF analysis. Near the end of the semester, a team of four students used a phase I Environmental Site Assessment (ESA) for the Alameda Beltway, a former railroad switching yard, as background and performed a screening study on the eastern portion of the Alameda Beltway site. Their research project utilized a modified version of EPA Method 6200 and the results indicated that lead contamination at greater than 200 mg/kg was observed for 47% of the sampling locations on the eastern portion of the Alameda Beltway making an in-ground community garden very impractical. The students' results were appended to the community partner's report to the city on the feasibility to plan for a community garden on the Alameda Beltway.

© 2014 American Chemical Society

Introduction

A service-learning field sampling experiment in an Environmental Chemistry course was centered on providing the community useful data on a specific environmental contaminant or pollutant. College and university chemistry departments typically have high quality instrumentation to investigate environmental contaminants and pollutants. A key factor for success is to adequately train students to perform an experiment to benefit the community. This training has many facets: having reliable background information on the site to develop a viable sampling plan; being proficient with the appropriate instrumentation; collecting representative samples and following a well-defined method of sample preparation and spectral data collection; and providing useful analysis of the resulting data in a professional manner. If a regulatory agency has formulated an analytical method, then the students can be trained to apply it and the regulatory agency's guidelines to interpret the resulting data yields an additional educational feature for the students. For students to recognize that the experimental data is valued by the community, they need to have good contact with the community partner.

A relationship built on reciprocity with the community partner is critical so that the research project can be successful as challenges or issues arise. This critical aspect is highlighted in numerous service-learning success stories (*1*). This reciprocity along with a compelling issue for the research project keeps students working diligently to provide the community high quality data and insightful analysis. In the research project discussed in this report, a strong reciprocal relationship between the community partner and the College existed, and the community partner approached the faculty member with their project which was directly aligned with previous screening studies performed in a collaborative fashion. Service-learning projects can yield tremendous learning experiences for students since the students recognize that their data is valuable to others and the issue driving the experimental work is indeed one that affects real people.

The compelling issue under investigation is the exposure to lead which is a known neurotoxin. Aspects of exposure to lead as a pollutant have been addressed in a few different service-learning experiments reported in the chemical literature (*1–3*). This issue is not novel, yet it is still compelling. Furthermore, there are sufficient resources so it readily functions as a terrific educational endeavor. In particular, the sources of exposure or the medium may guide the methods of analysis, such as lead in water may make the spectroscopic or electrochemical analyses more straightforward (*1–9*). A service learning project addressing lead paint was well documented in the *Journal of Chemical Education* (*2*) and the previously noted chapter in AAHE (*1*). In another *Journal of Chemical Education* article, this author focused on screening soil samples for lead at a residence known to have lead-based paint present (*8*). Plus, the screening of soils adjacent to a major urban highway was previously documented, and it is the training site that was used in this reported mini-research effort (*9*).

The XRF soil screening study of the Alameda Beltway site was a mini-research project presented to the instructor by our community partner, the Alameda Point Collaborative (APC). APC had received a contract from the city of

Alameda to evaluate the potential to establish a community garden on the Alameda Beltway site. The Alameda Beltway site is an abandoned railroad switchyard, and a phase I Environmental Site Assessment (ESA) was compiled in March 2010. One critical pollutant issue for the Beltway property was the presence of lead in the soils, however the report indicated that only a small number of soil samples were collected. Considering the property is approximately 24 acres in conjunction with the low sampling density in the ESA, additional soil sampling would be beneficial in characterizing the site. APC shared the phase I ESA with the instructor months before the beginning of the semester so planning the lab curriculum and allocating two weeks of lab to complete the mini-research project was straightforward to implement. The Alameda Beltway site is approximately 40 minutes from the campus which made it a bit more challenging logistically, however the opportunity to work with a terrific community partner, the Alameda Point Collaborative, to contribute to a community development project far outweighed this aspect.

The 2012 Environmental Chemistry course was therefore designed with a mini-research project as the final two weeks of lab. Students in this upper division course were allowed to select between two mini-research projects (a water quality project or a soil screening project). In both cases, an instructional lab was designed and implemented earlier in the course. The XRF soil screening instructional lab was integrated into the lab curriculum in 2002 and all students in the course participated in the instructional two-week lab. The students received instruction in the lecture on safe use of the X-ray emitting instrument. Students are provided an informed consent form with regard to whether they wish to trigger or not trigger the instrument. All students are required to process the resulting data from the instructional site, and there is no penalty if a student does not wish to collect a spectrum. The laboratory handout gave basics on preparing a location for spectral data collection and an industrial instrument video was used as a part of the instructional lab's preparatory materials. The instructional lab occurred on a highway median site with the appropriate permission from the State Department of Transportation (Cal-Trans). This field site is approximately 15 miles from the campus so transit to the site takes only a small portion of a 4-hour lab period. The instructional lab allowed students to study a site with a known pollutant source, which was the highway (deposition of lead from the combustion of leaded gasoline). Students received their training on how to establish a sampling grid on a field site and learned to work as a team. The students also became proficient using the field portable XRF instrumentation. Soil screening of the highway site introduced the students to U.S. EPA Method 6200 which is an XRF soil screening method (*10*).

The highway site has been used for several years and each time students do observe the same general trends for the lead in soil data. For each class of students, the experiment has plenty of new experiences when the class begins to establish a sampling grid using field tapes. Students must be observant that the site restricts their efforts to setup a perfectly uniform grid since the property is roughly triangular, the highway boundary is slightly curved, and the edges of the parcel were augmented with gravel and rock to provide proper drainage. These complications prepare the students for their subsequent mini-research study later in the semester. Furthermore, the students are not surprised that minor irregularities are likely to occur and that the minor adjustments need to be well documented in their notebooks. With some guidance by the instructor, the students make a series of measurements on lines that are approximately parallel to the highway and students computed an average soil lead at each location sampled. For a given parallel line, the soil lead values at each location are averaged and these values yield trends that are remarkably consistent except when the highway department moves the highway edge with new curbs. Fortunately collecting GPS readings when sampling began in 2002 has allowed subsequent classes to adequately address changes that occur at the site over several years and reference back to previous data sets (*9*).

Upon establishing a grid pattern on the field site, the students work in pairs to prepare each location following the *in situ* protocol of US EPA Method 6200. The students carefully uncover the surface soil with hand tools (shovel or hand trowel). Students clear the surface vegetation and remove any rocks and trash so that the surface soil can be homogenized using a hand trowel attempting to only pulverize the top few centimeters of soil. The area of exposed soil is approximately 100 square centimeters. This soil is packed in a slight dome shape using a wooden trowel so that the XRF instrument will lay flat on the surface without any obvious air gaps. Four experimental spectra are collected at different places of the dome-shaped 100 cm^2 area. The four spectral readings are recorded and the students are directed to compute an average lead content for the given location (*11*, *12*).

XRF Instrument Calibration

The U.S. EPA Method 6200 uses NIST soil standards which are recorded each day of operation at both the beginning and end of a sampling day. The Niton XRF 796 XLt manufacturer's manual recommends a warm up period of 30 minutes (to cool the detector to the appropriate temperature). An energy calibration of the X-ray tube is required each time the instrument is powered up to record quantitative data. This quantifies the X-ray flux of the instrument's source tube. To validate the detector energy counts, students record spectra of NIST standard reference material (SRM) soils 2710 and 2711 which represent a high and medium lead levels in soil, respectively. The instrument specifications delineate that the detector readings must be within 10 % of the NIST values or another energy calibration is necessary. The lead content in NIST SRM 2710 is 5532 ± 80 mg/kg and in NIST SRM 2711 it is 1162 ± 31 mg/kg of soil (*13*). The newer standards are NIST SRM 2710a at 5520 mg/kg and NIST 2711a at 1400 mg/kg of soil. To demonstrate instrument reproducibility and a lack of instrumental drift, spectral data of one SRM are recorded in triplicate at the beginning and end of the data collection. If the instrument is turned off or battery pack is exchanged the instrument is re-calibrated and the SRM samples are again recorded. All spectral readings have a 60-second minimum exposure time. At each typical field location, four spectra take approximately 5 minutes to measure.

Highway Instructional Site Results

The Environmental Chemistry students recorded spectral data after establishing a grid on the Highway 24 median. The class had three lines positioned at 30, 60, and 90 feet perpendicular to the highway. There were four locations on the line at 30 ft, four locations on the line at 60 ft, and three locations on the line at 90 ft. Plus two locations were sampled at 20 ft. The averaged lead in soil values for each distance from the highway are shown in Table 1. The general trend as a function of distance perpendicular from the highway is a steady decrease and these results match data previously recorded.

Table 1. Highway 24 Median Soils Analyzed for Lead by XRF

Lead in Soils (mg/kg)			
Distance from highway (ft)	Fall 2012	Fall 2006	Fall 2004
15	----------------	3122 ± 2586 (6)	2361 ± 1418 (7)
20	1107 ± 79 (2)	-------------------	-------------------
30	462 ± 214 (4)	637 ± 153 (3)	615 ± 185 (7)
60	338 ± 53 (4)	-------------------	-------------------
90	177 ± 35 (3)	-------------------	-------------------

The numerical lead content is listed along with the standard deviation of the values. The number in parentheses following the experimental values is the number of sampling locations at a given distance.

Alameda Beltway Site

The Alameda Beltway is an abandoned railroad line and the property has an industrial business park and residential areas surrounding it as shown in the Goggle Earth image below (Figure 1). The City of Alameda plans to have a public greenway park on the 24 acre site. The previous use as a railroad switchyard brings plenty of questions about contaminants being present. Since it is a relatively flat parcel, it is attractive for a portion to be developed as a community garden. The community food bank building is adjacent to the western end of the parcel, which could enhance the usefulness of the garden. However after a walking tour of the site with the community partner's representative, the decision to study the eastern portion of the site was made since it had less debris and future vehicle access appeared easier.

The eastern portion of the site is at least 2000 ft by 500 ft. There is a gravel road that partitions the site running east to west. Two sets of cement pads where railroad maintenance buildings once stood are present and adjacent to the road. The EIS indicates that petroleum residues and other contaminants were previously reported near the pads. The student field sampling plan focuses on the area south of the road which is open and reasonably level. Figure 2 shows a Goggle image of the site. Figure 3 is a schematic drawing of the site showing the sampling pattern. Additional sampling was conducted near the old railroad switch house since the building was most likely painted with lead-based paints. In Figure 2, the railroad switch house is located in the middle on the right near the Sherman St and it has been outlined in pale red. Two of the cement pads are outlined in the upper left of the photo in pale blue.

Figure 1. The Alameda Beltway in Alameda California; The Marina Square Business Park is to the north and residential areas are to the south of the Alameda Beltway. Map data © 2014 Google.

Figure 2. Eastern portion of the Alameda Beltway bounded by Sherman St. and Constitution Way. Map data © 2014 Google.

The orientation of the schematic diagram (Figure 3) has the eastern end of the site at the top. This representation matched the way that the students observed the site since the best entrance was at the far western end of the Alameda Beltway site.

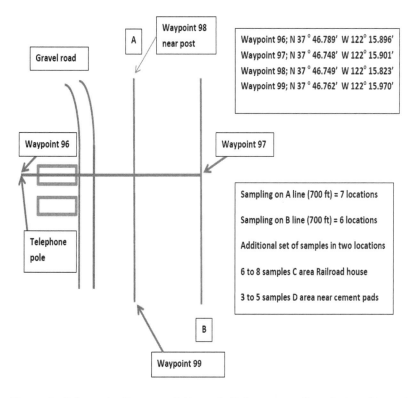

Figure 3. Schematic diagram of Alameda Beltway sampling design; Lines A and B are shown and GPS waypoints are noted on the figure. The cement pads are drawn on the left of the diagram.

The general composition of soils on line A and B were found to be highly variable from gravel with dirt to sand with dirt. The students organized sampling points along the A and B lines. The sampling locations were 50 ft apart on the respective lines and the B lines locations were staggered by 25 ft relative to the A line. In addition, one surprising feature was that many railroad ties were still on the site so this precluded easily establishing other parallel lines on the site. The railroad ties along the southern portion of the site were obvious once the student team walked the site, but the ties are not noticeable in the aerial view. Due the physical inhomogeneity of the samples along lines A and B, the student team decided that the soils near the old switchyard house should be sampled. These samples were labeled area C, and six samples were collected within three to five feet of the structure on the western and southern sides of the structure where exposed soils were accessible. As a final area of investigation, three samples were collected near the cement pads since the community partner thought the pads might be useful in setting up a garden supply storage area for the proposed future garden, and these samples were labeled as Area D.

Rainy weather occurred on the first day of field sampling so the student team set up their grid and collected damp soil samples in labeled plastic bags. To follow Method 6200, the soils were dried in the laboratory ovens for one week at 65 °C on aluminum foil trays. During the second week, students used mortars and pestles to grind the dried soils in the hoods. The soil samples were sieved using at least a # 60 mesh sieve. The resulting soil was placed in a new labeled plastic bag and spectral data were collected. Since the samples were homogenized by grinding and sieving, only three spectra were recorded for a given soil location. The lead values of the three spectra were averaged. If the spectra had been recorded in the field, the students would have recorded four spectra per location matching the training site soil screening *in situ* protocol.

Alameda Beltway XRF Spectral Results

The XRF spectral results of the NIST 2711 standard demonstrate the instrument was properly calibrated. The instrumental values on the NIST 2711 SRM were within 2% of the NIST listed value of 1162 mg/kg (*13*). The Alameda Beltway XRF spectral results ranged from a below detection limit to 875 mg of lead per kg of soil. The lead Lα and Lβ emission line intensities were quantified with the instrument software package. Two representative spectra are shown in Figure 4 and the lead Lα and Lβ emission lines are clearly identifiable at the energies of 10.55 and 12.61 KeV respectively. In addition, the representative XRF spectra shown have iron emission lines that are off scale. Copper and zinc emission lines are also identified in the spectra (copper Kα at 8.04 KeV and zinc Kα at 8.63 KeV). These elements were not quantified since the EPA preliminary remediation goals are considerably higher for these elements in comparison to lead, which was the focus of the screening study.

Reviewing the wide range of soil lead values, there were no obvious trends along lines A and B. The C area values were somewhat more consistent and the three soil lead values in D area nearest the cement pads had the highest lead value recorded on the site. The resulting lead in soil values are compiled in Table 2. As can be seen in the Table 2 for the 23 locations, 11 were determined to have soil lead values greater than 200 mg/kg.

Students evaluated the soil lead values using the US EPA preliminary remediation goals (PRGs) which set the residential soil limit at 400 mg/kg. Four locations had values that are greater than the EPA's residential PRGs. The lead in the soil was prevalent (~ 47% of the locations) as noted at a somewhat lower threshold of 200 mg/kg and the physical site survey demonstrated numerous issues beyond the concern with respect to lead in the soil.

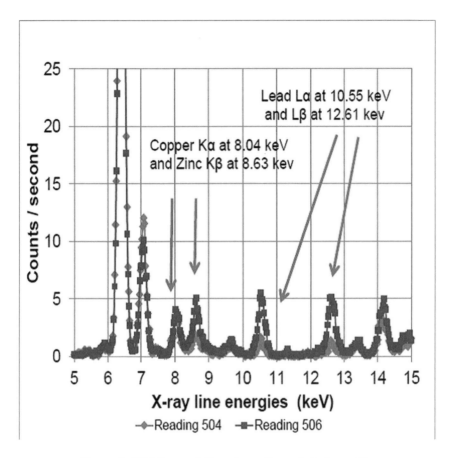

Figure 4. XRF Spectral Data from Alameda Beltway Site.

The irregularity of the lead content is also problematic since it does not allow the city to perform a remediation on one portion of the site and readily go forward with their plans to use the site as a park with an in-ground community garden. As noted, the other site features, such as the railroad ties still present throughout the site, pose additional hazards for garden construction and implementation. The rough gravel mixed with the soil and the obvious locations where illicit dumping has occurred also makes the site unsuitable for an in ground community garden due to the potential for other ill-defined contaminants. The students had some discussion with the APC representative whether a raised bed garden could be functional given the various issues on the Beltway site. The students' final conclusion relative to lead-in-soils was that a raised bed community garden could be planned for the Beltway site and this alternative would be sufficiently protective since the plants and the people growing the plants are not in direct contact with the contaminated soils. Whether the other limitations or issues would preclude a raised bed community garden was a decision that the students felt only the community partner, Alameda Point Collaborative, could finalize since APC had the expertise of running an urban farm in Alameda.

Table 2. XRF Soil Lead Values from Alameda Beltway Site

\multicolumn{6}{c}{AlamedaBeltway Samples Soil Lead}					
Location	Average Pb (mg/kg)	Std. Deviation	Location	Average Pb (mg/kg)	Std. Deviation
A1	Below Det.	----	C1	184	2
A2	316	13	C2	262	106
A3	718	96	C3	118	22
A4	402	57	C4	226	80
A5	117	12	C5	260	18
A6	484	39	C6	84	28
A7	54	11			
			D1	371	28
B1	96	9	D2	180	31
B2	137	7	D3	875	65
B3	146	9			
B4	359	14	NIST2711 Start	1136	25
B5	263	2			
B6	119	11	NIST2711 End	1135	26

Conclusions: Student Learning Outcomes

To assess the impact of this service-learning, mini-research project and the impact of the instructional labs, all enrolled students were requested to fill out a survey. The survey was created using the Student Assessment of Learning Gains tool (14). This survey tool has numerous standard questions focused on how various instructional tools help students learn the concepts. The additional utility of this survey tool is that faculty can create additional questions targeting specific educational aspects of a given course or lab. The survey questions can have options for the responses ranging from Yes/No, 5-point Likert scale to an open-text window. In this case, the instructor wrote a number of questions with respect to the initial XRF soil screening instructional lab and questions for the mini-research project. The overall class size was 9 students and only four participated in this soil screening research project, so statistical analyses are not relevant however the student responses and comments provide an instructor valuable feedback. For example, all students at the end of the semester expressed their thoughts on the XRF training experiment through an open-ended question. The question was what critical factor(s) could improve the data set quality for the highway 24 site training experiment. Two selected student responses were:

> "Several locations were not properly prepared [and similarly] the cleared area was not wide enough to allow for fresh soil to be measured for each [spectral] trial at each location."
>
> "I think if we had a chance to practice how to use the instrument prior to going out to the site, it would be beneficial, which includes preparing the selected soil spot to be x-rayed."
>
> In these two responses, it is clear students are expressing that adequate training with the instrument is necessary to obtain high quality data.

Since a smaller number of students answered the specific questions on their respective mini-research project which was placed at the end of the survey, their responses were very limited. However, a few key points were raised. The students expressed that spectral data collection would have be easier if the rain had not interfered and made the soils wet. Hence the students recognized that *in situ* spectral data collection was indeed a labor saving methodology. All the samples returned to the lab were always labeled so the students were also introduced to a chain-of-custody protocol. The students were divided in their opinions whether the site has limitations and whether the community partner could indeed safeguard future users of the site. After walking the site, the students were expressing concerns that additional hazards were likely present on the site. These additional hazards made the students wary about safe use of the site. In preparing their final powerpoint summary, the students questioned whether they should advocate that the site should not have any gardening or only raised bed gardening. The students were provided only limited information indicating how safe raised bed gardening could be if implemented properly. The presented final conclusion was that raised bed gardening would be an option for the community partner to evaluate since their expertise is in running an urban garden. The students' XRF results were included in the APC report to the City of Alameda, however it was submitted after the course was completed. It is the opinion of the instructor that if the students had additional instructional materials on raised bed gardening, their conclusion to support having raised bed gardens would have been voiced with more conviction. This aspect has given the faculty member insight on what additional materials need to be offered to students so they can be more confident in providing data and drawing conclusions to inform the public.

The mini-research concept allows students to observe their learning being put to action. One student even expressed the educational point of the mini-research project was they learned how our work can help the community. This salient point demonstrates that students do see the value of contributing to address community issues when labs are integrated to solve a community problem or provide evidence to assist a community in making a decision.

Acknowledgments

The Environmental Chemistry 2012 lab students are to be commended for working diligently on the various field sites. The students kept laboratory notebooks that were organized and with sufficient detail that the instructor could easily find the spectral data for each location of the field sites. The students worked to carefully report to the community what was observed. The students documented what resources were available from U.S. EPA to make an informed decision on the risks associated with contaminated soils.

This project was successful in thanks to our community partner, the Alameda Point Collaborative, and the City of Alameda. The no fee encroachment permit provided by Cal-Trans is acknowledged since the training site is also critical for the student training. The author acknowledges the critical feedback from colleagues in the SENCER project plus the encouragement from colleagues made him work to obtain the XRF instrument with the specific goal of including service-learning in the Chemistry curriculum. The Niton XRF elemental analyzer was purchase with the support of a grant from the Camille and Henry Dreyfus Foundation (SG 04-083). The instructor made presentations on this pedagogical work which were supported by the Saint Mary's College Faculty Development Fund and this funding is also acknowledged.

References

1. Giegengack, R.; Cressler, W.; Block, P.; Piesieski, J. In *Acting Locally: Concepts and Models for Service-learning in Environmental Studies*; Ward, H., Ed.; AAHE's Service learning in Disciplines; American Association of Higher Education: Sterling, VA, 1999; Vol. 6, pp 121–132.
2. Kesner, L.; Eyring, E. M. Service-Learning General Chemistry: Lead Paint Analyses. *J. Chem. Educ.* **1999**, *7*, 920–923.
3. Breslin, V. T.; Sanudo-Wilhelmy, S. A. The Lead Project, An Environmental Instrumental Analysis Case Study. *J. Chem. Educ.* **2001**, *12*, 1647–1651.
4. Weidenhamer, J. D. Circuit Board Analysis for Lead by Atomic Absorption Spectroscopy in a Course for Nonscience Majors. *J. Chem. Educ.* **2007**, *7*, 1165–1166.
5. Brouwer, H. Screening Technique for Lead and Cadmium in Toys and Other Materials Using Atomic Absorption Spectroscopy. *J. Chem. Educ.* **2005**, *4*, 611–612.
6. Wilburn, J. P.; Brown, K. L.; Cliffel, D. E. Mercury-Free Analysis of lead in Drinking Water by Anodic Stripping Voltammetry. *J. Chem. Educ.* **2007**, *2*, 312–314.
7. Goldcamp, M. J.; Underwood, M. N.; Cloud, J. L.; Harshman, S.; Ashley, K. An Environmentally Friendly Cost-Effective Determination of Lead in Environmental Samples Using Anodic Stripping Voltammetry. *J. Chem. Educ.* **2008**, *7*, 976–979.
8. Bachofer, S. J. Sampling the Soils Around a Residence Containing Lead-Based Paints: An X-ray Fluorescence Experiment. *J. Chem. Educ.* **2008**, *7*, 980–982.

9. Bachofer, S. J. Field Sampling with a FP-XRF: A Real World Lab Experience. *Spectrosc. Lett.* **2004**, *2*, 115–128.
10. U.S. Environmental Protection Agency. 6000 Series Methods; www.epa.gov/SW-846/pdfs/6200.pdf, Method 6200 (accessed August 2, 2014).
11. Shefsky, S. *Sample Handling Strategies for Accurate Lead-in-Soil Measurements in the Field and Laboratory* at the International Symposium of Field Screening Method for Hazardous Wastes and Toxic Chemicals, Las Vegas, January 29–31, 1997, www.clu-in.org/download/char/dataquality/sshefsky01.pdf (accessed August 2, 2014).
12. Shefsky, S. *Comparing Field Portable X-ray Fluorescence (XRF) to Laboratory Analysis of Heavy Metals in Soil* at the International Symposium of Field Screening Method for Hazardous Wastes and Toxic Chemicals, Las Vegas, January 29–31, 1997, www.clu-in.org/download/char/dataquality/sshefsky02.pdf (accessed August 2, 2014).
13. National Institute of Standards and Technology. *Standard Reference Materials: SRM*; https://www-s.nist.gov/srmors/certificates/archive/2710.pdf and https://www-s.nist.gov/srmors/certificates/archive/2711.pdf , No. 2710 and 2711 (Montana I and Montana II) (accessed August 2, 2014).
14. *Student Assessment of Learning Gains, (SALG)*; www.salgsite.org (accessed August 2, 2014).

Chapter 10

Bottled Water Analysis: A Tool For Service-Learning and Project-Based Learning

Olujide T. Akinbo*

Department of Chemistry, Butler University, Indianapolis, Indiana 46208
*E-mail: oakinbo@butler.edu.

Student's apathy toward learning and civic engagement is a global problem that is giving many governments, particularly in the developed nations, much concern. This is because of the potential adverse impact on their respective democracies. Based on recent data, it appears that apathy toward civic activities by contemporary youth is a way of expressing rejection of the traditional forms of civic engagement. Additionally it is also an expression of their distrust in the current political systems. On the other hand, apathy toward learning in general and science education in particular is a form by which the youths (students) are expressing their rejection of the practices in higher education. This includes both instructional method practices and the isolation of the academy from the society. In particular, students are quietly rejecting the notion of traditional instructional approach as an ineffective method. It is at odds with the way humans construct new knowledge. Additionally, isolation of the teaching-learning process from society denies students an opportunity to learn in context of the kind of problems and environments that they will have to engage upon graduation. It also denies them the opportunity to fully develop their power of creative problem solving. Education and society are inextricably connected; several scholars in the past and present have alluded to this fact. As such the apathies are interrelated at some level. To address these issues, we propose a hybrid intervention comprising of some elements of project-based learning and service-learning instructional methods. We have implemented this intervention several times with food safety monitoring as the focus. In this chapter, we

© 2014 American Chemical Society

present the general frame work of the intervention, the logistics and results of an implementation based on drinking water analysis, the impact on students (based on self-reported data), faculty and community. Based on observation and analysis of data collected in an end-of-semester course evaluation, it can be inferred that students engaged the teaching-learning process and accomplished the intended learning outcomes in content and skill development. Finally, we provided guidance for potential adopters and adapters.

Introduction

Science education has undergone many reforms since its recognition as a distinct area of education in the 19th century (*1, 2*). Yager (*3*) noted about 40 major educational reforms in the 150 years of US history. Tobin and co-workers classified recent reforms into four groups or movements (*4*) each with its own trigger and driver. Policies were put in place and interventions were developed to address root causes and overcome challenges presented by each trigger. One trigger that is currently and commonly reported is apathy toward science disciplines, particularly in the laboratory. A coinciding trigger alongside apathy toward science is apathy toward civic engagement. This has also been reported by others from various segments of the society. Combined, these two apathies have become the current drivers of reform in science education. This chapter is focused on a proposed hybrid intervention (Project-based and Service-Learning instructional method) that was implemented to address these drivers. This was used for the laboratory experience in an introductory analytical chemistry course. Project-based learning provided the framework, service-learning provided the context and food analysis (in this case water quality) provided the focus for the experience. The chapter is intended to provide the basis/rationale for the proposed hybrid intervention by highlighting the root causes of the problems and addressing common concerns. Furthermore, it presents a description of an implementation focused on water quality (as an example of other focuses that have been utilized) and offers advice for potential adopters/adapters, both to allay common concerns and provide guidance for success.

Apathy: A Major Problem in Science Education and Civic Engagement

Apathy toward Learning, Contributing Factors, and Its Global Impact

Based on the purpose and focus of this chapter, apathy toward learning can be described as negative attitudes such as lack of active involvement, commitment and attention in the classroom and/or laboratory. It is characterized by irregular attendance, non-participation in class discussions, sleeping in class, engaging other activities (e.g., texting, doing other course homework or studying for another course's test) while the class is in progress, not doing the home work or doing the barest minimum to get by. In the laboratory, it can be demonstrated by thoughtlessly following a procedure as a recipe to generate data without purpose or with the purpose of just fulfilling the requirement for the time allotted. Apathy reflects students' lack of motivation and interest in the subject matter or current exercise.

Apathy in the science classrooms and laboratories has been reported (at times implicitly) in several books, journal articles and reports that focus on active learning. They are too numerous for this chapter to even fully capture. To get a sense of it, the reader is directed to read any literature on inquiry-based, guided-inquiry, problem-based, discovery-based, cooperative, collaborative, POGIL and case based instructional methods. All of these were developed in part to address the lack of enthusiasm in classroom and laboratory and to facilitate the development of several skills among which critical thinking is common. However, a few classics are recommended here for reading (*5–8*). Apathy in the classroom has been demonstrated at various levels of learning from K12 through higher education. The following story is a sample experience reported by a K12 teacher:

> *"During this past year, I have observed students in my mathematics classroom who needed to take more initiative toward their education. They needed to complete more assignments and participate more in class. There was not just one student who demonstrated the above behavior, but there were many. For example, [John Doe] was a prime example of an apathetic student. He had failed the previous grade and was in danger of failing for three quarters in a row. [John Doe] turned assignments in late, and he missed many assignments. He also did not participate in class well. He decided instead to socialize and to disrupt the class. The sad fact is that there are many students in America like [John Doe] who are apathetic toward their education. Why are these students apathetic and what can be done to decrease the number of students who have this care free attitude toward their education? What is the cause of student apathy (9)?"*

Unfortunately, attitude and interests along with several other factors are known to be predictors of performance in any discipline particularly science (*10*). Reynolds and co-worker proposed nine major factors that affect student achievements in the sciences. These are classified into three groups. The first group comprise ability, motivation, and effort. The second group consists of instruction related factors including quality and quantity. The third group includes social and psychological environments (*11*).

A major consequence of apathy toward learning science is the decreasing number of science majors particularly at the higher education level (*12*). According to Wang and co-worker, in spite of the investments and effort in STEM education, the number and gender composition of graduates continues to fall short of demand (*13*). Jones and co-worker also noted the increasing gap between the number of science graduates needed to meet environmental, technological and economic needs in the society and the number of students choosing to pursue the study of sciences (*14*). Additionally, they alluded to world-wide occurrence of the problem and the call for ways to stimulate students' interests in science disciplines.

Apathy toward Civic Engagement

Apathy toward civic engagement describes observations such as the public's disinterest in civic and political affairs, public's distrust in democratic process, and the current attitude of individualism (or the notion of not being responsible for or accountable to each other) that pervades both societal and academic culture today. Apathy toward civic engagement has been observed in a variety of forms. The following story provides an example of the "individualism" mentioned above.

"I have been desensitized over the years with news stories of human suffering and tragedy. In my academic studies in criminology, I have read, encountered, and studied both the purely bad and truly evil acts that humans have done, and through this, I have built up a resistance from these deviants, and often criminal cases. However, once in a while, a case pops up that adds a new level to this playing field that supersedes my definition of sick and twisted. The news story of two year old Wang Yueyue who was run over twice on October 13th, 2011, and then eight passers-by walked by her as she lay in the back alley injured from the first vehicle impact (CBCnews, ABC news & BBC news Television, October 14th to October 21, 2011). All of this was captured on a security video camera. .... My point to this post is why so many people walked by while the little girl lay on the road at the point of death. Perhaps, as many have said on the media, we must change our values and take action. Some have gone further and saying that laws need to be rewritten to reflect a high moral standard, while others are saying that the law is not the answer, but teaching people to be proactive rather than being non-active (15).

Manifestations

In 1995, Putnam noted that there was a decline in voting, civic participation, and public trust (*16*). In 1998 the National Commission on Civic Renewal (NCCR) reported a decline in the quality and quantity of public voluntary service (*17*). Demographic data indicated that this decline was more pronounced among the youth (*18*). In 2000, Lane among other scholars noted that the decline in civic engagement phenomenon was not localized to USA alone but is widespread among western societies (*19*). In 2005, the *Millennium Development Goals* of the United Nations indicated that it is a global phenomenon (*20*). In 2007, Bellah and co-workers (*21*) warned that the individualism tendencies in the culture were threatening the traditional social bonds in the American society. Colby and Ehrlich (*22*) further commented that the growing sense of individualism dominates the culture at the expense of a broader social, moral and spiritual meaning. The ultimate result of such attitude is a fragmented and polarized society. However, there is a school of thought that disagrees with the notion of declining youth engagement in political and civic activities. Instead, they believe that these youth are rejecting the tradtional forms of participation and interelations and are emerging with newer forms of engagement.This thought has been well captured in the works of Stolle and coworkers (23). Youniss and coworkers highlighted that this is a global phenomenon too (24). Data from the Cooperative Institutional Research Program (CIRP, an arm of the Higher Education Research Institute, University of California, Los Angeles) seems to corroborate this view. CIRP collects data annually from entering freshmen to assess their development of civic-responsibility sense through college years. These students were compared to past students based on two factors: first, involvement in volunteerism and community service and second, interest in politics. Data indicated that these cohorts volunteered more for civic activities but were more politically disengaged. Furthermore, the results indicated that as they progressed in their college experience, these students became more committed to helping others in difficulty, influencing social values, influencing political structures and participating in community action programs. However, many of these gains disappeared in few years after college. Scholars agree that these result indicated that students are civically engaged but are placing their efforts in areas where they believe it can make a difference (25).

Root Causes

How does this relate to the academy? Unfortunately some scholars have identified the culture and practice in higher education as a contributing factor of apathy toward civic engagement. This blame comes partly from the traditional instructional style that pervades the academy. According to Colby and Ryan, this approach helps in guiding students to examine and analyze phenomena but is ineffective in helping them to anchor their own experiences in moral and civic lessons from complex texts. In other words students are not allowed to develop their own framework. In his criticism, Zlotskowski (26) claimed that the practices and learning outcomes of the universities have not changed in over a century. The traditional instructional method predominantly used in the classroom is neither adequate for developing civically engaged citizens nor is it adequate for producing graduates that meet needs of the society. Moreover, it is at odds with the learning style of majority of students. Additionally, there is also the tension between the way the academy sees its mission as that of developing expertise and human capita. As such, any activity that does not contribute to expertise is undervalued and underfunded. These situations combined to trigger bad experiences for some students. The matter is further complicated with the tendencies of some instructors' to misinterpret the differences in the learning preferences of students as deficiencies and lack of intellectual ability. Thereby turning what could have been a trigger of interest and motivation into a liability (27, 28).

Impact

The impact of this practice and position is the creation of unengaged students. Palmer (*29*) pointed out that 'Every way of knowing [of learning] tends to become a way of living'. Unfortunately the current way of learning in the academy [and the culture of the academy] breeds competitive individualism that pushes people to make objects of each other and see the world as means to be exploited for private gains. This according to Zlotkowski (*26*) can only lead to a depersonalized society. The author went on to note that, if disengaged students can evolve into to disengaged citizens, even engaged students whose experiences are limited to their own academic achievements can wind up as citizens who have never learned to see their fellow citizen or the society at large as more than an object to be exploited for their own personal achivements.

Remedy

There have been many calls to higher education to renew its role in the strengthening of the American democracy by re-socializing people (*30*). Checkoway (*31*) noted that many American universities were established with a mission to prepare students for active participation in democracy and to develop knowledge for the improvement of communities. However, commitment to this charge has waned among university administrators. Harkavy (*32*) suggested that the goal for universities should be to contribute significantly to developing and sustaining democratic schools, communities and societies. To accomplish this goal, he suggested that higher education should effectively educate students to be democratic, creative, caring, and be constructive citizens of a democratic society. Furthermore, Harkavy submitted that "When colleges and universities give very high priority to actively solving strategic, real world problems in their local community, a much greater likelihood exists that they will significantly advance citizenship, social justice and the public good". Abbot and Ryan suggested that if young people are to be prepared and equipped effectively to meet the challenges of the 21st century, it is advisable to base it on the best understanding of how humans learn. They also recommended that learning should be integrated into larger community where students can engage real issues and and responsibilities (*27*). Earlier in 1991, Moon (*33*) suggested that contemporary liberal education must pay attention to the ways that knowledge might be used in practical ways and must connect academic content to the development of civic goals. According to Colby and Ehrlich (*22*), a morally and civically engaged individual would recognize himself or herself as part of a larger society and would see social problems as his or her own. These authors also suggested that the enhancing of moral and civic engagement will require a combination of knowledge, virtue and skill. This cannot be accomplished by reading a single book or taking a single course. Instead, each book and each course contributes its own quota to the development of such an attitude.

According to Cassidy (*34*) each individual has his/her own way of approaching learning (i.e., each individual has his/her own learning/cognitive style, approach or strategy). Examples of learning styles include: instructional preference, social interaction, information processing, cognitive personality, wholist analytic, personality centered, cognitive centered, and activity centered. The author provided further overview on each model. Abott and co-worker (*35*) further stated, that there is a global need for new competencies; the ability to conceptualize and solve problems that entail abstraction (manipulation of thoughts and patterns) and require system (or interrelated) thinking, experimentation, and collaboration beyond what can truly be fully delivered in the classroom. According to him, higher order thinking and problem solving skills grow out of direct experience. The same sentiment was echoed by Stenberg (*36, 37*) when he suggested that the practice of using well-structured problems to teach is a disservice to the student given the fact that most real world problems are ill structured. The strategies that work for those hypothetical problems may fail in real life. As such, the best preparation for solving real life issues can only be made in the community with real life, ill-structured problems. These suggestions, recommendations and criticisms along with several others resulted in the activities that are now commonly referred to as service learning. Several universities have incorporated service-learning into their programs. For example the "Campus Compact is an association of more than 500 college and university presidents who are committed to service-learning (*31*)". In addition colleges and universities are fostering partnerships with their communities to facilitate civic engagements for their students (*38*).

Rationalization of Project-Based and Service-Learning as a Two-Prong Intervention for Addressing Apathy toward Learning and Apathy toward Civic Engagement

Confluence of the Two Apathies: Interrelationship of Science, Science Education, and Social Responsibility

Science and society are inextricably tied together. Perhaps this is the reason why apathy toward science education and civic engagements are also interrelated. It is clear from previous sections that a poorly prepared, unengaged student may graduate to be civically unengaged as well. In other words, the two apathies are intertwined and can influence each other. This interrelationship has also been recognized since the inception of science as a field of study in the 19th century. According to DeBoer, (*39*) the inclusion of science education into the school curriculum was justified on the basis of its relevance to contemporary life and its contribution to a shared understanding of the world by all. The writings of scholars such as John Dewey, Thomas Huxley, Herbert Spencer, Charles Lyell, Michael Faraday, John Tyndall and Charles Elliot were significantly instrumental in this process. Of course these were advocates of constructivism. Also, in 1918, a report of the Commission on Reorganization of Secondary Education (an arm of the National Education Association, NEA) (*40*) underscored the interrelationship between science and social responsibility as reflected in the following comments:

"Importance of applying knowledge: Subject values and teaching methods must be tested in terms of the laws of learning *and application of knowledge to activities of life* [emphasis mine] rather than primarily in terms of the demands of any subject as a logically organized science"
"The purpose of democracy is so to organize society that each member may develop his personality primarily through activities designed for the well-being of his fellow members and the society as a whole ....Consequently, education in a democracy, both within and without the school, should develop in each individual the knowledge, interests, ideals, habits and powers whereby he will find his place and use that place to shape both himself and the society toward ever nobler ends"

It seems whatever affects one will affect the other. Therefore it is not surprising to note that literature on panacea for both apathies always track back to the works of John Dewey, Jean Piaget and the progressives who all hold the constructivist view of education practices.

Confirming the interrelationship of education and social responsibility further, one of the four scholarships (scholarships of discovery, integration application and teaching) that Boyer (*41*) presented in his work "*Scholarship reconsidered: Priorities of the Professoriate*" is scholarship of application. This scholarship ties learning/knowledge to social responsibility as well. According to the author, this is the scholarship where the following questions are asked: "How can knowledge be responsibly applied to consequential problems?"; "How can it be helpful to individuals as well as institutions?"; "can social problems themselves define an agenda for scholarly investigation?" He went on to remind his readers that the higher education system was founded on the principle of serving the interest of the larger community. But then he immediately highlighted the widening gap between the values in the academy and the needs of the larger world. Boyer also advised that scholarship of application does not mean that knowledge always precedes application, but that knowledge can be gained during application (i.e., during service). These scholarship categories work interdependently with each other.

However, Freire (*42, 43*) implicated instructional practices in higher education as a root cause of apathy toward civic responsibilities. In his scathing review, the author characterized traditional instructional approach that pervades the academy as oppressive and dehumanizing. He used the metaphor of *Banking Education"* to describe the method. In this model, the teacher acts as a bank clerk and students are the depositories where the clerk files away his currency (knowledge). As such the students are just vessels for storage who regurgitate back what is stored in them at the behest of the teacher (i.e., during assessment). This thought is better captured in the following excerpts from the article:

"The capability of the banking education to minimize or annul the students' creative power and to stimulate their credulity serves the interest of the oppressors, who care neither to have the world revealed nor to see it transformed"

"Unfortunately, those who espouse the cause of liberation are themselves surrounded and influenced by the climate which generates the banking concept, and often do not perceive its true significance or *its dehumanizing power* [emphasis mine]".

"The "humanism" of the banking approach masks the efforts to turn women and men into automatons – the very negation of their ontological vocation to be fully human"

The author implies that current practice does not give students enough power to transform their minds and hence their reality. In this regard he wrote:

"Those truly committed to liberation must reject the banking concept in its entirety, adopting instead a concept of women and men as conscious beings, and consciousness as consciousness intent upon the world. They must abandon the educational goal of deposit-making and replace it with the posing of the problems of human beings in their relations to the world"

"Indeed, the interest of the oppressors lie in "changing the consciousness of the oppressed, not the situation which oppresses them"; for the more the oppressed can be led to adapt to that situation, the more easily they can be dominated"

"The banking approach to adult education, for example, will never propose to students that they critically consider reality. It wills instead with deal such vital"

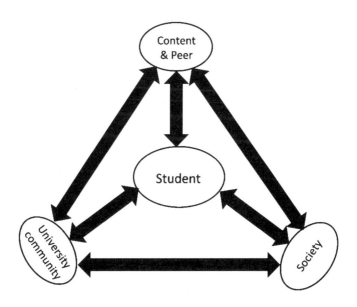

Figure 1. The interrelationships that must happen among student, university community, society and content for student to be ready to solve societal problems.

Implicit in his view is the fact that a dehumanized student is a dehumanized citizen. A dehumanized citizen has no capacity to think for himself, be creative nor is he/she likely to engage transformation of his world. Conversely, if the student is allowed to flourish in in his consciousness, he will grow to become a creative citizen with capacity to engage. Boyer (*44*) in 1996 recognized that the higher education had lost its position in the scheme of things and recommended a way back. The way back is embeded in what he called the scholarship of engagement. According to him "scholarship of engagement means connecting the rich resources of the university to our most pressing social, civic, and ethical problems, to our children, to our schools, to our cities......"

Therefore, any intervention therefore must strongly consider the interrelationships among the following components: students, university community, society and content (see Figure 1).

Given all of the above, it is clear that the solution to overturn the apathies can probably be implemented simultaneously in the same context or setting. Additionally, such a solution will have to start with changes to the way that teaching and learning is conducted in higher education.

The Panacea

With their apathy toward learning, the students are rejecting the traditional instructional approach. Additionally with their apparent apathy toward civic engagement the youths are rejecting the traditional forms of civic engagement. Therefore it is obvious that contemporary adolescents are yearning for a different mode of operation. In the words of Harkavy (*45*), "When colleges and universities give very high priority to actively solving strategic, real world problems in their local community, a much greater likelihood exists that they will significantly advance citizenship, social justice and the public good". Based on these, we are therefore proposing a hybrid intervention comprised of some elements of project-based learning (an inquiry based instructional method) and service-learning (to ground the work with a focus that benefits the society). The two instructional methods are further discussed next.

Project-Based Learning as an Option for Facilitating Knowledge Acquisition, Skill Development, and Attitude Changes Needed To Address 21st Century Challenges

There are several instructional methods ranging from those that are primarily lecture based (or teacher-centered) to those that are primarily activity based (or student-centered). These instructional methods are classified based on the extent of instructor's involvement in their implementation (*46, 47*). However, selecting one of them for purpose requires that one should take into consideration current understanding on how people learn and the skills that are needed to meet contemporary human challenges. These are commonly referred to as the 21st century skills. Although there isn't a formal list, McComas (*48*) reported one list of 21st century skills which includes: "life skills (agility, flexibility, and adaptability), workforce skills (collaboration, leadership initiative, and responsibility), applied skills (accessing and analyzing information, effective communication, and determining alternative solutions to problems), personal skills (curiosity, imagination, critical thinking, and problem solving), interpersonal skills (cooperation and teamwork), and non-cognitive skills (managing feelings)". Obviously, in addition to these skills there is the need for knowledge of content as well. The science laboratory is a unique place to teach and learn these skills (*49*). Some goals that can be accomplished in well-designed laboratory experiences include: enhancement of students understanding of concepts and their applications; development of scientific practical skills; and problem-solving abilities; development of scientific "habits of the mind"; understanding of how science and scientists work; and stimulation of interest and motivation in the sciences (*50*). Several authors have recommended active learning as the instructional method to accomplish these goals (*7*). Woods (*51*) catalogued a list of 33 variants of active learning instructional methods and used the degree to which students are empowered to identify the subtle difference among them. In addition, he provided a method for selecting an effective learning method. In our case we selected the project-based learning for its efficacy in facilitating content reinforcement and skill building.

What then is project-based learning and how is it different from other active learning instructional methods? How is it implemented and what are the benefits? According to Bell, *"project-based learning is a student driven, teacher facilitated approach to learning. Learners pursue knowledge by asking questions [inquiry] that have piqued their natural curiosity"* (*52*, *53*). Furthermore, Bell described project –based learning as a framework for teaching students through investigations. In this framework, students pursue solutions to authentic problems through designing of plans, inquiry and refining of ideas, debating of ideas, making predictions and testing of ideas, collections and analysis of data, getting information from data, drawing conclusions, communication of findings to others, asking new questions and creating products. Ideas for the project may come from the students or the instructor. However, students must have a hand in its development and the outcome of the project must not be constrained to a pre-determined result. It is somewhat difficult to distinguish project-based learning from several other active-learning instructional methods that preceded it and even those that were innovated after it. Thomas (*54*) recognized this problem and provided the following criteria for characterizing and/or identifying project-based learning:

1. The project utilized must be central to the curriculum and teaching strategy. Student should be able to learn the central concepts of the discipline through the project. Essentially the project is the curriculum.
2. The project's driving question (s) must present sufficient challenge for students particularly in the area of concepts and principles of the discipline. The project could be based on themes and/or interdisciplinary topics. The activity should have important intellectual purpose with potential for contributing new knowledge.

3. The project must be have a clearly defined goal and should be implemented with an investigative approach facilitates development of new skills, new knowledge and opportunity to reflect and draw some conclusion from the results and activity.
4. Students must experience a certain degree of independence in designing, implementing and making sense of the results (or information generated). In other words students must experience a level of autonomy.
5. The project must be focused on a real-world issue and must be authentic. It must not be a simulated experience with a predetermined outcome.

Thomas is not the only one to highlight the difficulty of defining the uniqueness of project-based learning. Larmer (55) attempted to distinguish between project-based learning, problem-based learning, and fifteen other X – based learning instructional approaches. According to the author, project-based and problem-based learning are similar conceptually. They both utilize open-ended questions or tasks, facilitate application of content and skills, helps learners to develop 21st century skills, emphasize student independence and inquiry. Additionally they both require longer time for implementation compared to traditional or classical approach. However, project-based learning is often multidisciplinary, requires a longer time for implementation, and typically results in a final product (data, or performance). Also, it is often focused on real world problems in natural settings and contexts. On the other hand, Problem-based learning, is often single discipline based, and requires relatively shorter time for implementation. It also involves the utilization of fictitious scenarios and the outcome is a proposed solution to the problem.

Regardless of the way it is defined, project-based learning is mostly used to motivate students to engage the teaching –learning process, to facilitate skill building and changes in habits of the mind. It is used to provide a research-like experience for students. For some, this is their introduction to research and for others, this is their only research or research-like opportunity in their college career. Santiago and coworkers (56, 57) reported that students who were exposed to mixed-method comprising of cooperative and problem based lab instructions demonstrated improved problem solving and strategy. Additionally, these students also showed an increase in their ability to regulate their metacognitive skill in spite of reduced guidance. Based on a review of research on project-based learning, Thomas (54) concluded that the approach is effective for delivering the 21st century skills.

With project-based learning, project tasks are similar to professional realities and require more time (days, weeks) to accomplish goals rather than the canned 4-hour experiments that are utilized in traditional laboratory settings). Projects are more directed toward application of knowledge rather than rote or robotic following of procedures. With projects, planning, and personal initiatives (seeking guidance, seeking information and appropriate resources needed for success) time and resource management are crucial if any progress will be made toward accomplishing the set goals. Project-based learning has been implemented both in the classroom and in the lab (55–70) and can also be used to teach an entire course (*71–73*). In this case the laboratory and classroom experiences are combined into one activity (*74*).

The next section is used to describe and rationalize the choice of Service-learning (the other component of the instructional approach that we adopted).

Service-Learning as a Tool for Motivating Civic Engagement

The service-learning instructional approach is used in our context as a motivational tool to create awareness of civic responsibility while learning at the same time. It is also used to overcome the social apathy that was described earlier in the opening paragraph. Just like project-based-learning, service-learning can be defined in a variety of ways. According to Bringle and Hatcher (*75, 76*), "*service-learning is defined as a course based, credit –bearing educational experience in which students (a) participate in an organized service activity that meets identified community needs and (b) reflect on the service activity in such a way as to gain further understanding of course content, a broader appreciation of the discipline and an enhanced sense of civic responsibility*" Saitta and coworkers (*77*) defined service-learning as "*involving students in activities that serve the community but also connects those activities to learning goals for a course.*" In principle, service-learning can be summarized as "learning by serving and doing". In service-learning all parties involved (the student, community and instructor) benefit. The instructor gains improved content delivery and collaboration opportunities with peers and community. The community gains valued services while the students gain course –related skills, increased self-confidence and broadened sense of community engagement (*78*). Historically, teaching in a way that emphasizes giving back to the community is not new. Like project-based learning, the principle of learning via experience can be traced back to the work of John Dewey (*79–81*) and other education philosophers who promoted the idea that the fundamental goal of higher education is to prepare citizens for community service (*82, 83*). However, literature on 'service-learning' as an instructional tool are relatively new (*84*). Most of this literature started coming out in the late 1990s'. Service-learning as an instructional tool was designed in part to address the observed disconnection between the academy and the community for which it produces its graduates. In 1993, Astin noted that the culture of American universities was dominated by materialism, individualism and competitiveness. These values adversely affected the universities themselves and also contributed to the fracturing of the sense of community in the nation (*85*). Between 1997 and 2002, the American Association for Higher Education published an 18 monograph series titled "Service-learning in the Disciplines". One of the monographs focused on service-learning in Biology (*86*). The articles in this twelve chapter monograph demonstrated the effectiveness of service-learning as an instructional method that helped to build a bridge between the academy and the community. Eight of the chapters were narratives of experiences in various courses from a variety of schools. All seem to suggest good experience for all partners involved (the academy, community students and instructors). Writing on the impact of service-learning on biology curricula, Kennell (*87*) claimed that service-learning enhanced learning, provided real-world experiences, empowered both students and teachers, facilitated student-centered courses, and civic responsibility in the society. Sutheimer (*88*) claimed that service-learning also facilitated improved cognition for students.

As an instructional method, service-learning can be implemented with a variety of models. The three most dominant of these models include: the philanthropic, the civic and the communitarian models. The philanthropic model assumes that the job of the university is to help people in acquiring the intellectual capability to solve contemporary problems through development of discipline-based skills and of moral values (*89*). However, opponents of this view claim that the philanthropic model creates a social stratification of the 'philanthropists' and the 'needy'. They also claimed that it is founded on the ideas of noblesse oblige, responsibility, privilege and dependence. According to Sementelli (*90*), these ideas undermine many of the objectives of service-learning. It makes the "*....students to serve because they are more fortunate socially and economically than those receiving the service...*" As a result students may see the activity as another requirement rather than a context for building civic and cultural awareness. The philanthropic model also, conflicts with one of the cardinal desired outcomes of higher education – critical thinking. The civic engagement model views higher education as an agent of civic renewal with social change as one of its goal. This is the most commonly used model of service-learning (*91*). According to Hollander (*92*) this model became quite prominent in the 1990s' as a result of the perception that civic engagement by Americans was waning. In 1998 the National Commission on Civic Renewal (NCCR) reported a decline in the quality and quantity of public voluntary service (*93*). This was indicated in –part by the decline observed in voting, civic participation, and public trust (*16*). As result, there was a call to Higher education to renew its role in the fortification of the American democracy by re-socializing people (*30*). In fact, Checkoway (*31*) noted that many American universities were established with a mission to prepare students for active participation in democracy and to develop knowledge for the improvement of communities. However, a commitment to this mission has waned among university administrators. Harkavy (*32*) echoed the same sentiment by suggesting that universities should be to contribute significantly to developing and sustaining democratic schools, communities and societies. He further advocated that higher education should effectively educate students to be democratic, creative, caring, and be constructive citizens of a democratic society. He also submitted that "When colleges and universities give very high priority to actively solving strategic, real world, problems in their local community, a much greater likelihood exists that they will significantly advance citizenship, social justice and the public good". In response to these calls, several universities have incorporated service-learning into their programs. Additionally, Campus Compact, an association of college and university presidents who are committed to service learning has been formed (*31*). Also, more colleges and universities are fostering partnership with their communities to facilitate civic engagements for their students (*38*). The Communitarian model sees the students and classroom as part of a larger society (*94*). The students are members of the communities of their discipline. As such they must be trained in the professional and ethical skill of the community. This community of professionals transcends national boundaries (see Figure 2).

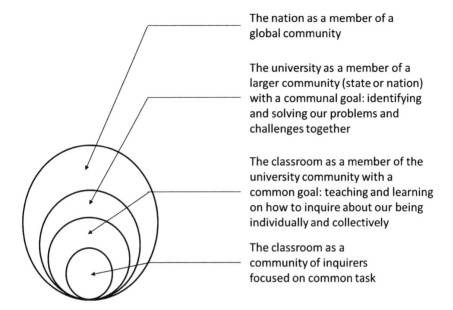

Figure 2. *The learner as a member of a global community.*

Reports on outcomes of service-learning implementation are replete in literature. However, a few are highlighted here. Kammler and co-workers (*78*) reported service-learning project that was focused on environmental monitoring of water samples collected from a nature reserve (Glen Helen). The uniqueness of this project lies in the fact that it is a collaboration between two colleges. This is similar to the work of Hatcher-Skeers and Aragon in which the collaboration was between college and middle school students (*95*). It also facilitated mentoring of freshmen undergraduate students by graduate and upper-class students. In addition, two separate courses were involved. Students had hand-son experience on EPA methods 200.7 to determine the metal contaminants of the water, and 300.1 to determine the anion content. Also, students carried out microbiological analysis of the samples. Students developed a standard operating procedure and disseminated their results through oral presentations. Based on these, it is obvious that the students experienced a full range of science practitioners experience. According to these authors, research indicates that service-learning benefits every party involved. Students develop course relevant skills, enhanced self-confidence, and improved sense of community engagement. The instructors gain better content delivery experience, and collaboration with both peers and community. The community gained free service and interaction with university community. In the next section the results of an implementation of the project-based, service-learning is discussed.

Table 1. Summary of Content and Skill Intended for Each Project

Project	Content Addressed	Skill intended
Writing workshop	None	• Communication through technical report-writing • Application of spreadsheet for large data analysis and data reduction • Application of word processor for scientific report generation. • Function, content and format of component parts of technical articles
Data analysis	Titrimetric techniques, stoichiometry Q or G, t, F-test and ANOVA	• Statistical analysis and reduction of data • Presentation of data using figures and tables • Collaboration • Result communication • Technical report writing
Method Comparison	Calibration approaches, electrochemical, spectroscopic, chromatographic techniques theory and application (students can consult ASDL site (http://home.asdlib.org/) for introduction to techniques not yet covered at the time.) Statistical evaluation of data and methods	•Statistical analysis and reduction of data • Presentation of data using figures and tables • Collaboration • Result communication • Technical report writing
Method development	Calibration curves, Quality Assurance (figures of merit, data quality objectives, quality control during chemical analysis)	•Quality assurance • Method optimization • Statistical analysis and reduction of data • Presentation of data using figures and tables • Collaboration • Result communication o Technical report writing

Capstone Project	Sampling, statistical analysis, calibration curves, figures of merit (as applied to method and instrument), data analysis and data reduction,	•Project planning,
		• Experiments design,
		• application of concepts,
		• Collaboration
		• Result dissemination & communication
		  o poster preparation and presentation,
		  o Technical report writing

Preliminary Implementation Hybrid Intervention: Project Based-Service-Learning Experience in Analytical Chemistry 1

The Framework

A framework for teaching analytical chemistry at Butler University was first reported in 2008 (*96*) and was further described in 2013 by Akinbo (*97*). In this framework, Analytical Chemistry 1 (or Quantitative Analytical Chemistry) is taught as a single course comprising of lecture and lab components. Previously used experiments are now replaced with coherent projects that build up to a capstone exercise. With this approach, students are provided with the opportunity to address an authentic real-world problem by applying the knowledge and skills they have acquired throughout the semester. The content and skill areas for each of the projects are summarized in Table 1. Some project reports are written collaboratively in groups, while others are written individually. Reports of the final capstone project are typically presented as posters which are prepared collaboratively in groups. Assignment of report type varies for projects annually. In the case of Analytical Chemistry 2 (or Instrumental Analysis), the lab and lecture are now separate independent courses. The lab courses are theme-based and entirely project driven. The project has an overarching question that is divided into smaller questions. The weekly lab exercise focuses on the sub-questions. Two examples of themes that have been implemented are environmental analysis, and food analysis. Additional themes have been utilized by other faculty members.

This framework has been applied to study herbal supplements in which we investigated potential exposure to trace element contaminations due to the varied geographical sources and ability of plants to preferentially accumulate some metals. We have also studied fish and fruit juices to compare domestic and imported samples based on trace elements. We have studied soils to investigate the suitability of urban soil for gardening of edibles. In this chapter, the results of bottled water analysis will be used to demonstrate the implementation of the capstone-project portion of the framework.

Sample Capstone Project from Analytical Chemistry 1: *Bottled Water Analysis*

The Project Background

Below is a quote from the opening paragraph of one of the student groups (*98*):

"Bottled *water is a growing billion dollar industry all over the world. As the industry grows, regulation of bottled waters and other available drinking waters for the public has come into question (Ikem and co-workers (99)). The health costs versus the benefits of drinking such water has been weighed many times, and on one hand there are claims that certain minerals have medicinal and therapeutic effects (Kermanshahi and co-workers (100)) while others have found that there may be components that interfere with endocrine activity and produce health hazards (Plotan and co-workers (101)). Nonetheless, it is important to regulate the amounts of trace elements and minerals in bottled water to ensure the safety of the public."*

The Project's Overarching Question and Goals

The major (overarching) question of this project is "*is drinking bottled-water worth buying?*" The goal is to get students to use their knowledge and skills to address a real-world issue. It is also desired that the students will experience project planning, experimental design, acquisition of reliable data based on good laboratory and quality assurance practices. Other goals include having hands-on experience on statistical analysis of data, data reduction, interpretation and presentation. It was also intended that students will use information obtained from their data to arrive at personal decisions with regards to quality of bottled water and how it compares to tap water. In the process, students will enhance their chemical analysis and team-work (or collaborative) skills.

At the end of the process students will have valid information to choose one bottled water brand over the other. Also, the project will help the students to understand certain concepts more, and also build on the skills (technical and social) that they have previously developed. The impact to the society will come as students' communicate their findings to their friends and family and to the audience that attend their poster presentation.

The Logistics

Thirty water samples were analyzed for 30 elements. These comprise of four mineral elements (K, Ca, Na, Mg) and 26 trace elements (B, Be, Al, Sb, As, Ba, Be, B, Cd, Cr, Co, Cu, Eu, Ho, La, Pb, Mn, Mo, Ni, Se, Ag, Sr, Tl, Th, U, V). Only the mineral elements are presented in this report. The sample set comprise of, 8 purified (PW), 6 sparkling (SW), 6 natural spring water (NSW), 1 artesian (ART) 1 distilled (DW), 6 municipal tap water (MTW), and 2 well water (WW) samples. The WW and one set of the MTW were obtained from homes of a staff and faculty member respectively. There were three types of the MTW: untreated (U), softened (S), and softened-filtered (SF).

An inductively coupled plasma mass spectrometry (ICPMS) was used for the analysis. The class worked in groups comprising of 3-4 students. At the front end, each student read and summarized two relevant journal articles of their choice. Members of each group merged their summaries into only one document that became the introduction section of their report. Each group came up with a sampling and sample preparation plan that was discussed as a class. A sample list was compiled and a sample preparation standard operating procedure (SOP) was developed that the entire class used. Each student group was responsible for analyzing 6 samples. During analysis, data was collected for instrument and method performance characteristics for purposes of quality control. Additionally, data quality objectives (DQO) were set to determine acceptability of data (see Table 2).

Table 2. Data Quality Objectives (DQO)

Quality Parameter	Quality Objective
ISTD Recovery Stability	100 ± 30%
% Recovery from Spiked Sample	100 ± 30%
Accuracy of CRM	100 ± 30%
Precision	≤5% RSD
Continuous Calibration Verification (CCV)	100 ± 30%
Relative Percent Difference (RPD)	≤ ± 30%
Correlation coefficient for calibration plot (r2)	≥0.99
Contamination tracking (CCB or blank verification)	≤20% of initial value

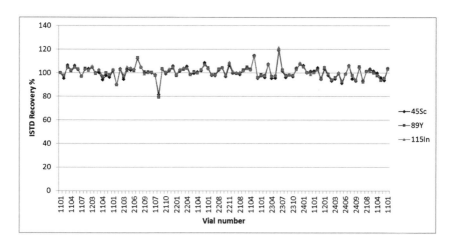

Figure 3. Instrument stability monitoring through internal standard signal tracking.

The Project Results

Instrument Performance Characteristics

Instrument stability was monitored during each analysis batch by tracking the signals of the internal standards, ISTD (Sc, Y and In) for each sample. The signal for each ISTD in each sample was divided to the first signal acquired in that batch. The ratio was then multiplied by 100 to obtain the parameter called ISTD Recovery %. An example plot of the stability data is presented in Figure 3. Generally the instrument performed within specification (80-120 % ISTD recovery) throughout the duration of the batch analysis.

Other instrument performance characteristics (calibration parameters, detection limit, background equivalent concentrations, and correlation coefficients) are presented in Table 3.

Table 3. Instrument Calibration Parameters, Detection, Limit, and Quantitation Limit for Each Mineral Element

Isotope	Correlation coefficient	Slope	y-intercept	Detection limit (ppb)	Background equivalent concentration (ppb)
23 Na	0.9990	0.00535	0.0826	1.10	15.4
24 Mg	0.9997	0.00272	0.00164	0.090	0.605
39 K	0.9989	0.00221	0.233	5.94	105.8
43 Ca	0.9992	0.00004	0.00003	1.86	0.945
44 Ca	0.9998	0.00038	0.00097	1.15	2.54

To track instrument calibration stability, the 50 ppb calibration standard was measured at intervals between samples. The result is presented in Figure 4. The result indicates that the instrument calibration remained generally stable during the batch analysis, save for a period when the potassium, K, signal dropped to about 50%.

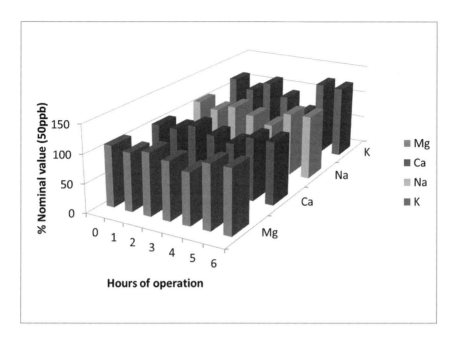

Figure 4. Instrument calibration stability check. This was done with the 50 ppb calibration standard solution. (see color insert)

Instrument accuracy was checked with a certified reference material of trace metal in drinking water (CRM-TMDW-B), which was purchased from High Purity Standards. The CRM was re-analyzed periodically in-between group of samples to ascertain instrument accuracy stability. The results are presented in Figure 5. The percent accuracy for all elements met the desired data quality objectives

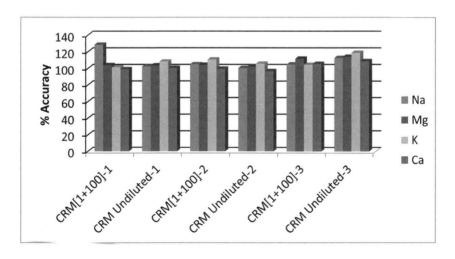

Figure 5. Instrument accuracy using CRM-TMDW-B. (see color insert)

To check instrument reproducibility for the analyte, in the sample matrix, a small set of samples were analyzed in duplicate. Relative percent difference (RPD) was calculated for each pair and the results are presented in Table 4. The data quality objective was met for most of the samples except for the PW which failed for all elements. This could be specific to this one sample. Also, the RPD's were high for potassium, K.

Table 4. Relative Percent Difference (RPD) of Select Samples (One Brand Per Type of Water Sample) Analyzed in Duplicate. This Was Calculated as the Ratio of the Difference between the Duplicates to Their Average. The Ratio Was Multiplied by 100. This Was Used To Check Reproducibility of the Results.

Sample Type code	Sample type	Relative Percent Difference (RPD)			
		23Na	24Mg	39K	44Ca
NSW	Natural Spring Water	19.2	14.4	74.3	14.8
SW	Sparkling water	-1.4	-2.3	27.8	-0.5
PW	Purified water	71.6	86.0	44.8	134.8
CRM	Certified reference material	-5.4	-4.4	-29.2	-8.6
CRM	Certified reference material	-1.7	-2.3	-3.7	-10.0
MTW	Municipal tap water	1.7	0.1	-8.5	6.0

Results obtained for the water samples are presented in Table 5 and Table 6.

Comparison of Bottled Water, Tap Water, and Well Water Samples

From results presented in Table 5 and Table 6, it is apparent that bottled water samples generally have lower concentrations of magnesium (0.2-7.6 ppm Mg) compared to the municipal tap water (29 -81 ppm Mg) and well water (31 ppm Mg) except for sparkling water sample 2 (With 52 ppm Mg) and natural spring water 1 (NSW-1 with 31 ppm Mg).

Also, the untreated municipal tap water and untreated well water generally contained a higher concentration of Na (61-81 ppm for MTW and 31 for WW) compared to the bottled water samples except in the cases of sparkling water sample 1, 2, and 6 (SW1, SW2 and SW6) (see Table 6).

Table 5. Concentrations of Mineral Elements in Bottled Water Samples

Bottled water Sample	Sample Code	23Na		24Mg		39K		44Ca	
		Conc (ppm)	RSD	Conc (ppm)	RSD	Conc (ppm)	RSD	Conc (ppm)	RSD
Natural Spring Water	NSW-1	9.2	21.6	33.2	17.5	3.8	60.0	19.2	18.4
Natural Spring water	NSW-2	8.8	4.1	7.6	4.8	2.5	18.5	11.1	5.4
Natural Spring water	NSW-3	9.0	1.6	27.1	1.5	1.8	10.8	22.5	1.8
Natural Spring Water	NSW-4	7.8	2.3	6.9	1.3	1.6	20.3	6.4	2.6
Spring Water	NSW-5	5.0	2.7	3.5	2.8	1.2	23.2	2.8	5.1
Natural Spring water + Electrolytes	NSW+E	1.6	9.0	0.1	13.0	0.04	15.4	2.0	8.0
Sparkling Water	SW-1	119.4	0.9	0.9	1.3	1.2	2.1	1.4	5.7
Sparkling Water	SW-2	35.2	3.1	52.8	2.9	2.6	13.4	55.9	2.4
Sparkling Water	SW-3	14.4	10.0	6.4	8.8	2.2	48.7	33.2	9.6
Sparkling Water	SW-4	10.7	2.2	4.4	0.3	1.0	19.0	44.3	1.8
Sparkling Water	SW-5	6.8	1.4	2.2	0.3	0.7	22.7	3.4	4.8
Sparkling Water (With Sodium Added)	SW-6	31.2	2.6	9.4	2.1	3.6	4.7	10.9	1.8
Artesian	ART	20.4	1.7	16.7	0.8	6.4	6.7	6.6	6.4
Purified Drinking Water Sodium Free	PW-NOSO	5.5	0.7	0.1	9.4	3.2	5.9	0.4	8.1

Continued on next page.

Table 5. (Continued). Concentrations of Mineral Elements in Bottled Water Samples

Bottled water Sample	Sample Code	23Na		24Mg		39K		44Ca	
		Conc (ppm)	RSD	Conc (ppm)	RSD	Conc (ppm)	RSD	Conc (ppm)	RSD
Purified Water	PW-1	4.5	15.2	2.7	12.9	2.6	55.4	1.9	12.2
Purified Water Enhanced with minerals for pure fresh taste	PW-2	6.9	4.2	3.5	5.3	5.8	9.8	0.3	18.6
Purified Drinking Water	PW-3	7.0	5.3	0.2	7.4	3.7	14.8	0.5	9.1
Purified Water	PW-4	5.0	1.7	0.6	5.5	8.5	3.2	1.5	14.3
Purified Water	PW-5	4.6	1.3	6.9	4.0	21.3	3.2	4.6	2.9
Purified Water	PW-6	13.8	1.5	3.8	1.1	2.7	15.1	4.9	4.0
Distilled Water	DW	5.2	2.7	0.2	7.7	2.4	11.7	0.3	19.7

Table 6. Concentrations of Mineral Elements in Municipal Tap and Well Water Samples

Sample Description	Sample Code	23Na Conc. (ppm)	23Na RSD	24 Mg Conc. (ppm)	24 Mg RSD	39 K Conc. (ppm)	39 K RSD	44Ca Conc. (ppm)	44Ca RSD
Municipal Tap Water From Faucet	MTW-B-L	61.8	1.2	29.3	1.4	5.2	4.2	23.6	0.6
Reverse Osmosis Cleaned Butler Municipal Tap Water	MTW-RO	6.1	3.2	0.1	8.6	0.6	24.2	0.5	24.8
Municipal Tap Water From Fountain	MTW-BF	74.3	1.1	29.6	2.5	6.6	4.8	27.0	0.4
Municipal Tap Water From Fountain Duplicate	MTW-BF-D	73.1	2.5	29.5	1.8	7.2	4.6	25.4	1.8
Municipal Tap Water From Fountain with Filter	MTW-BF-F	39.3	15.3	23.9	15.2	2.3	81.0	19.3	16.1
Well Water untreated	WW-JW-U	31.3	4.2	39.8	3.7	3.9	12.3	24.5	3.0
Well Water softened	WW-JW-S	214.1	0.9	0.2	12.3	1.8	8.6	3.3	9.7
Well Water softened and filtered	WW-JW-S-F	219.3	1.7	0.8	6.1	2.0	9.3	1.3	13.5
Municipal Tap Water Un-softened	AK-MTW-U	81.2	7.6	34.4	7.4	7.9	16.3	22.1	7.0

Continued on next page.

Table 6. (Continued). Concentrations of Mineral Elements in Municipal Tap and Well Water Samples

Sample Description	Sample Code	23Na		24 Mg		39 K	44Ca	
		Conc. (ppm)	RSD	Conc. (ppm)	RSD	Conc. (ppm)	RSD	Conc. (ppm)	RSD
Municipal Tap Water Softened	AK-MTW-S	48.2	8.6	20.5	7.8	4.1	26.6	13.0	7.9
Municipal Tap Water Softened and Filtered	AK-MTW-SF	65.6	13.7	26.4	12.5	7.0	31.1	18.1	11.1

Based on Tables 5 and 6, it can be inferred concentrations of potassium are fairly comparable in both sets of sample ranging from 0.04 -8.5 ppm K. Except for one bottled water sample, purified water sample 5 (PW-5) with concentration of 21.3 ppm K. The sparkling water samples generally have higher calcium concentrations (1.4 -55.9 ppm Ca) than other bottled water samples. The purified water samples have the lowest concentrations of calcium (0.3-6.6 ppm Ca). The concentrations of Ca in the tap water (23-27 ppm Ca) and well water (24.5 ppm Ca) are more similar to the natural spring water samples (see Table 5).

Municipal Tap Water from Campus

Municipal tap water samples MTW-B-L and MTW-RO are from the same location on campus. Comparing their results shows that the reverse osmosis (RO) system reduced concentrations of the mineral elements significantly by removing 90% of Na, 99% of Mg, 89% of K and 98% of Ca. The fountain water sample (MTW-BF) contained higher Na and Ca concentrations compared to the sample from the faucet. It can therefore be inferred that the fountain bled some Na and Ca into the water. However, attaching a filter to the fountain resulted in a reduction of the mineral elements by 47% of Na, 9% of Mg, 65% of K and 28% of Ca 65% .

Municipal Tap Water from Faculty Member's Home

A second set of municipal tap water samples (AK-MTW-xxx) was collected from a faculty member's house located about 12 miles from campus. Softening reduced the mineral elements in these samples by 41-48% with the highest reduction occurring for K (48%). However, filtering did not seem to reduce the mineral elements. The faculty member acknowledged that the filtration cartridge on the refrigerator from which the sample was collected needed replacement.

Well Water from Staff Member Living in a Suburb

The well water samples were collected from the house of a staff member who lives in a sub-urban area. Comparing the results of the untreated well water sample (WW-JW-U) with the softened well water sample (WW-JW-S), it can be concluded that the water softener reduced the mineral elements by 99% for Mg, 54% for K and 87% for Ca. The result for Na is inconclusive due to an outlying data obtained for the untreated sample which might be due to a glitch during data acquisition. Also, comparing results of softened (WW-JW-S) and filtered-softened water samples (WW-JW-S-F), it appears that the water filter in this household was not removing the minerals. As such the filter needed replacement.

Making Sense of the Data Using Principal Component Analysis

Principal component analysis was used to further interpret the data. The results from this analysis are presented next.

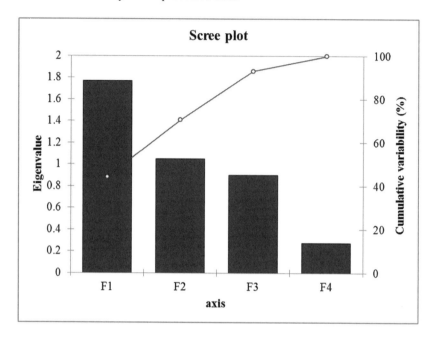

Figure 6. Contribution of Factors to variability of samples.

Based on the Scree plot (see Figure 6), the first three factors account for most of the variability among the drinking bottled water samples. Combination of factor 1 (F1), and factor 2 (F2) alone already accounts for 70.5% of the variability. Adding factor 3 (F3) to this raises it up to 93.1%. Correlation between factors and the variables (i.e., the mineral elements) is presented in Table 7.

Table 7. Correlations between Variables and Factors

	F1	*F2*	*F3*	*F4*
23Na	0.239	-0.701	0.672	0.026
24Mg	0.886	0.269	0.107	-0.362
39K	-0.268	0.727	0.626	0.090
44Ca	0.914	0.135	-0.095	0.371

As shown on the table, Factor 1 (F1) correlates strongly with Mg and Ca. These two minerals typically relate to hardness of water. Therefore, it is inferred that F1 is a measure of water hardness and it discriminates among the samples based on their hardness. Factors 2 and 3 (F2 and F3) correlate with Na and K. These two elements contribute significantly to the taste of water. However F2 distinguishes between Na taste and K tastes while F3 is unable to distinguish between them. This was inferred based on the observation that Na correlates negatively with F2 while K correlates positively with it. Both Na and K correlate positively with F3. Together, both F1 and F2 (i.e., hardness and taste) account for about 71.48% of the variability among the samples (see Figure 7). This suggests that hardness and taste are the most distinguishing factors among these water samples. A biplot of sample-factor (samples correlations with F1 and F2) in Figure 8 shows the relative hardness and relative taste of the drinking bottled water samples. On this plot it is evident that the purified water samples are relatively softer than the other samples. On the other hand the artesian and natural spring water samples are relatively hard. The taste of most of the spring water (except NSW1, NSW3) and most purified water (except PW2, PW4 and PW5) are sodium controlled (see Figure 9).

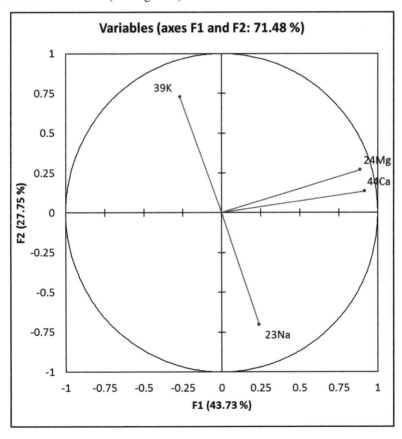

Figure 7. Factor-variable correlations.

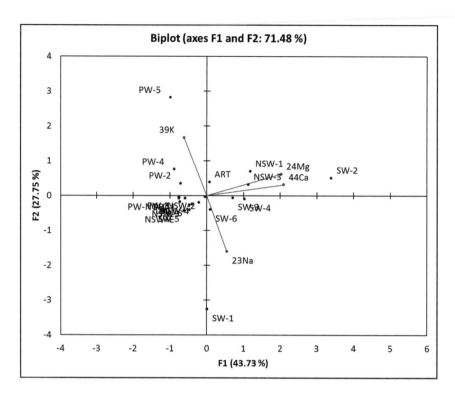

Figure 8. Sample–factor correlations. Distribution of sample mostly by hardness (F1).

Combining the Results of all the Water Samples (Bottled, Municipal, and Well Water Samples)

To reduce the complexity of the combined data, a Varimax rotation was done to enhance data visualization and interpretation. Factor loadings after the varimax rotation indicates that dimensions, 1, 2 and 3 (D1, D2, and D3) cumulatively accounts for 94% of the variability in the samples (see Table 8). Again D1 alone accounts for 44% of the variability and correlates with Ca and Mg. This indicates that dimension 1 relates to water hardness. Also, it indicates that hardness is the most distinguishing factor among the samples. Dimensions 2 (D2) is strongly correlated to K while dimension 3 (D3) correlates strongly to Na. D1 and 2 are not correlated to each other.

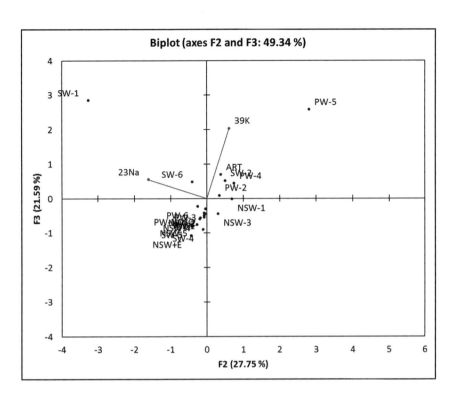

Figure 9. Sample–factor correlations. Distribution of sample mostly by taste (F2).

Table 8. Correlations between Variables and Factors after Varimax Rotation

	D1	D2	D3
23Na	0.008	-0.030	**0.999**
24 Mg	**0.924**	0.175	0.054
39 K	0.044	**0.995**	-0.031
44Ca	**0.941**	-0.081	-0.036

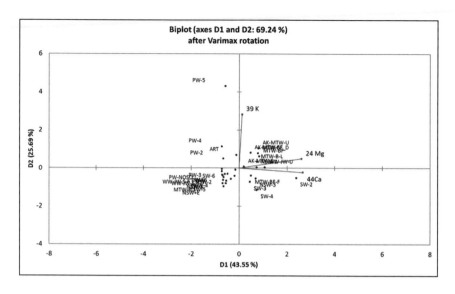

Figure 10. Biplot of variables and factors after Varimax rotation.

The biplots of dimensions 1 and 2 after varimax rotation (Figure 10): reveals the classification of all the samples into four clusters. The D1 axis, correlates with Ca and Mg which can be inferred as water hardness dimension. The samples that fall into the harder water clusters are: all the softened municipal tap water samples (AK-MTW—U, MTW-B-L), softened municipal tap water (AK-MTW-S), softened and /or filtered municipal water (AK-MTW-SF, MTW-BF, MTW-BF-D, MTW-BF-F), untreated well water (WW-JW-U), some natural spring water (NSW-1, NSW-3) and some sparkling water samples (SW-2, SW-3, AND SW-4). All the purified water samples, the artesian water, the distilled water and Sparkling water 1 (SW-1, SW-5, 0, softened and /or filtered well water, reverse osmosis treated municipal tap water (MTW-RO) and distilled water (DW) were classified as less hard . In summary, the municipal tap water and well water are harder than most of the bottled water samples. Notice that we did not infer that the bottled water samples are necessarily soft.

Conclusion on Drinking Water Analysis

Based on all the data obtained, it can be concluded that the major difference between municipal tap water, well water and drinking bottled water is determined by two major factors, hardness and taste. The municipal tap water samples that we analyzed in this study were generally harder than the bottled water. The intervention and project provided positive experiences for all the parties involved (students, community and instructor)

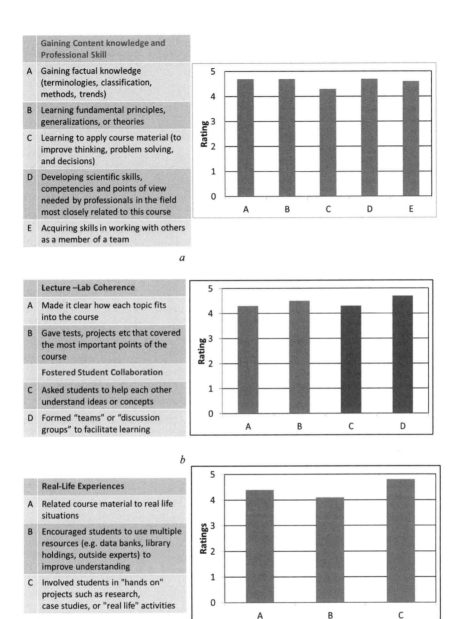

Figure 11. Results of student assessment of their learning gains and teaching effectiveness (a) gains in content knowledge and skill area, (b) coherence of lecture and laboratory activities and (c) experiences with real-world issues. Note: these groupings are not typical IDEA grouping. These were done by the author to address issues relevant to the focus of this chapter.

Results and Reflections on Impacts of Implementation of the Hybrid Intervention on Students, Faculty, and Community

Impact on Students

The impact of the project experience on students was garnered from course assessment data. This was conducted by using a forty three-question survey provided by IDEA Company (a nonprofit organization that provides evaluation and development services for institutions. See: http://ideaedu.org/services/student-ratings). The questions were then sub-categorized into the three groups presented in Figures 11a, 11b and 11c. Figure 11a was a check of student perspective on their own gains in the area of content knowledge and professional skill development. Students scored this category above 4 on a 5-point scale indicating that they believed that they made significant progress in this area. Next the perspective of students on the coherence of the lecture and lab was probed. Again, students scored this category generally above 4 on a 5 point scale thus indicating that students were able to see the connection between concepts that were covered in the classroom and the applications in the laboratory. When students were probed for their view of team-work experience (collaboration), they rated it above 4 on a 5-point scale indicating that they recognized that the course implementation facilitated the building of this social skill for them. Finally, we probed student perspective on whether the course facilitated learning for them using real-life, hands-on experiences. Again they responded with an average rating above 4 on a 5-point scale indicating that this is true. Based on these results, one may infer that the instructional method applied (which in our case was a combination of cooperative learning-lecture in the classroom, and project-based learning –service-learning in the lab were effective. Students acknowledged that they gained content knowledge, professional skills, and real-life experiences. The results also suggested that that the lab and lecture were coherent (not aligned topic-for topic), and that the instructional approaches fostered collaborations among the students.

Impact on Community

In our case we recognize the campus as a community of its own albeit part of a larger community. As such the community that we have served mostly is the Butler community. We have plans to expand our activities to the immediate community to which the campus belongs, particularly the Tarkington area community of Indianapolis, Indiana. With regards to the impact of this project on the community, a staff-member and a faculty member received free water analysis. The results that they received included both the mineral and trace elements as well. Additionally, the effectiveness of their water softener and water filtration systems was communicated to them. The larger campus community also benefited through the presentations. One was done by the instructor during a campus-wide event (Butler University Brown bag Series) and another was done by the students during the annual department of chemistry poster session event. It is also envisioned that students will communicate the outcome of their work to their peers' and family.

Impact on Instructor

The project result has facilitated a professional development opportunity for the instructor via publication of a book chapter (this one) and presentations on campus and at professional meetings. This work was presented at the recently concluded *Biennial Conference on Chemical Education (BCCE 2014)*. Additionally, the results from the project constitute the focus of this chapter. This also facilitated for the instructor an opportunity to demonstrate (1) to the administrators on campus the outcome of their investments in a recently purchased ICPMS and (2) to the university community of our capability and the impacts that chemistry department is making in the lives of our students.

Recommendation for Implementers

There are thoughts that this author will like to leave with the reader and potential adopter of the intervention.

Avoiding Mistakes of the Early Implementers

Project-based learning principles are not new; they date back to the days of Dewey (*79, 81*). However, they were not widely adopted initially for variety of reasons. According to Blumenfeld (*53*), early implementers underestimated

1. What it took to motivate and engage students in complex work.
2. The depth of knowledge and skill required of students in designing and executing a project.
3. The demands on the instructor in terms of knowledge, skill and commitment (time, energy and resources) pedagogical content knowledge (i.e., illustrations, representations, analogies, examples, and explanations to make the subject matter comprehensible to the students).
4. The complexity of classroom organization associated with open ended projects (e.g., risk-taking); dealing with unexpected results and seeing it as an opportunity for learning and not project failure; breaking down of questions to executable experiments; and linking concept taught in the classroom to the experiments.
5. The resource requirements (e.g., technology) for successful implementation.

Handling Failed Experiment: Use Unexpected Results as an Occasion for Troubleshooting and Learning Not as a failure

One of the reasons why some resist moving away from using the traditional experiments in teaching science laboratory courses is the perception that open-ended laboratory projects are prone to fail. This implies that a successful laboratory experience is one that results in a predetermined or expected result. This is contrary to the real-world experience of professional scientists. Moreover, it is known that the traditional experiments do motivate students to engage science. How then should success be defined in this context and how does an instructor plan for success? First of all, In the National Science Education Standards (NSE) of 1996 (*102*) science was defined as "a way of knowing that is characterized by empirical criteria, logical argument, and skeptical review". In other word science is based on inquiries upon the natural world or systems. Scientists probe and the system reveals its characteristics. Scientists do not impose on nature pre-determined characteristic. The NSE further recommends that "Students should develop an understanding of what science is, what science is not, what science can and cannot do, and how science contributes to culture". Bell (*103*) also recommended that laboratory activities should be designed to pique the interest of the learner. A successful laboratory experience will have the following qualities:

1. The project should have a meaningful focus/question that is based on real-world issue which is of interest to the student.
2. Students should be involved in the project-design/plan and the plan should derive from literature review of appropriate and relevant information.
3. Students should implement the thought-out plan religiously. However, the plan must be flexible and open to modification if preliminary results suggest doing so.
4. The project facilitates opportunities for students to test their hypothesis, generate and analyze data, and then make decisions based on information obtained from the carefully generated data.
5. The outcome of the project experience is disseminated in an appropriate forum in writing, orally or via poster presentation to an appropriate audience.
6. The project should facilitate the development or enhancement of desired skills (e.g., 21st century skills),
7. The project should facilitate in-depth inquiry,
8. The project should involve peer-peer and instructor-learner critiquing of report with revisions.

Most of these ideas were adapted from the report of Larmer and Mergendoller (*104*).

On the issue of instructors' role to ensure successful implementation, Blumenfield and co-workers (*53*) provided a guide. The instructor:

1. creates opportunity for learning by providing access to information
2. provides instruction and guidance to make tasks manageable
3. stimulates students to use learning and metacognitive processes,
4. assesses student progress, troubleshoots problems, evaluates results and provides feedback to students
5. facilitates an environment that is conducive to constructive inquiry
6. Manages the classroom/laboratory to ensure orderly and efficient project implementation

However, an implementation can fail if:

1. The project is planned on a resource that is not available or known to malfunction
2. The project does not have the financial support of the administration
3. Student presents a lackadaisical attitude toward the project, thus leading to
    a. Compromised samples and standards
    b. Careless operation of instrument
    c. Improper interpretation of results
4. The instructor lacks expertise on the planned project and has no support from a knowledgeable source
5. The students and/or instructor are not willing to invest time in troubleshooting of minor issues and hence gives up in frustration.

These are things to consider prior to engaging the implementation of this approach.

Considering Major Benefits of Project-Based Learning: Research Experience

Benefits of the project-based learning and service-learning instructional methods have been mentioned in preceding sections. However, one major benefit not previously addressed is the fact that these two approaches facilitate research experiences for undergraduate students who would not have had such opportunity in their tenure as an undergraduate. Literature is replete with benefits of undergraduate research experience (*105–109*). Moreover, there have been calls for the introduction of research experiences into science courses as well (*110, 111*).

Conclusion

In conclusion, there is a growing concern about the decline in student attitudes toward science education and toward civic engagement. The problem is global and it has caused many governments to commision studies that have led to a variety of reports, recommendations and policy changes. In this chapter, we have described the individual apathies, identified their root causes and highlighted their interrelationships. We also presented a brief review of project-based learning and service-learning methods. These instructional methods have been successfully used independently to facilitate inquiry-driven learning and to connect learning with the real world. A hybrid instructional method that is based on these principles was presented in this chapter as an option for addressing problems of apathies toward learning and civic engagement. In this regard, a drinking water analysis project was utilized to engage students in inquiry activities, enhance knowledge acquisition, develop chemical analysis skills, build 21st century skills and render service to the campus community. The rational for selection of instructional methods, the results and impact of their implementation were also presented. Based on student evaluation of their experience it can be concluded that the intervention was effective. Additionally, highlights of potential pitfalls and advice on successful planning were provided for potential adopter/adapters.

Acknowledgments

The author will like to acknowledge Barbara Howes of the science library at Butler University for her support in securing articles and interlibrary loans. The author will also like to acknowledge Oreoluwa I. Akinbo for proof-reading the manuscript.

References

1. DeBoer, G. E. *J. Res. Sci. Teach.* **2000**, *37* (6), 582–601.
2. DeBoer, G. Historical Perspectives on Inquiry Teaching In Schools. In *Scientific Inquiry and Nature of Science: Implications for Teaching, Learning and Teacher Education*; Flick, L. B., Lederman, N. G., Eds.; Springer Academic Publishers: Netherlands, 2004; pp 17–35.
3. Yager, R. E. *The Clearing House* **2000**, *74* (1), 51–54.
4. Tobin, K.; Tippins, D. J.; Gallard, A. J. Research on Instructional Strategies for Teaching Science. In *Handbook of Research on Science Teaching and Learning*; Gabel, D. L., Ed.; Macmillan Publishing Company: New York, 1994; pp 45–93.
5. Bonwell, C. C.; Eison, J. A. *Active Learning: Creating Excitement in the Classroom*; The George Washington University: Washington, DC, 1991.
6. Woods, R. W. *Ind. Eng. Chem. Res.* **2014**, *53*, 5337–5354.
7. *Active Learning: Models from the Analytical Sciences*; ACS Symposium Series 970; Mabrouk, P. A., Ed.; American Chemical Society: Washington, DC. New York, 2007; p xii, 291 p.
8. Kuwana, T. *Curricular Developments in the Analytical Chemistry*; A report from the workshops. 2009; pp 1-50; http://www.asdlib.org/files/curricularDevelopment_report.pdf (accessed Aug 04, 2014).
9. Cheney, L. *Student Apathy - How to motivate and drive students by self-paced learning*; 2010. Schools-Teachers-Parents: Issues in schooling Website;

http://schoolsteachersparents.wikidot.com/student-apathy (accessed Aug 04, 2014).

10. Singh, K.; Granville, M.; Dika, S. *J. Educ. Res.* **2002**, *95*, 323–332.
11. Reynolds, A. J.; Walberg, H. J. *J. Edu. Psychol.* **1991**, *83*, 97–107.
12. Osborne, J. *Int. J. Sci. Educ.* **2003**, *25*, 1049–1079.
13. Wang, M.; Degol *J. Dev. Rev.* **2013**, *33* (4), 304–340.
14. Jones, K. P; Barmby, P. *Int. J. Sci. Educ.* **2007**, *29*, 871–893.
15. Hall T. *Social Apathy, or a Psychological Problem*, 2011; Thomasso's weblog, http://www.thomasso.com/2011/10/29/social-apathy-or-a-psychological-problem/#sthash.qBywi2qZ.dpuf (accessed Aug 04, 2014).
16. Putnam, R. D. *J. Democracy* **1995**, *6*, 68.
17. Bennett, W. J.; Nunn, S. *A nation of spectators: How civic disengagement weakens America and what we can do about it*; National Commission on Civic Renewal; University of Maryland: College Park, MD, 1998.
18. Colby, A.; Ehrlich, T.; Beaumont, E.; Rosner, J.; Stephens, J. Higher Education and the Development of Civic Responsibility. In *Civic Responsibility and Higher education*; Ehrlich, T., Ed.; The American Council on Education and Oryx Press: Westport, CT, 2000; pp xii−xliii.
19. Lane, R. E. *The loss of happiness in market democracies*. Yale University Press: New Haven, CT, 2000.
20. United Nations. Department of Public Information. *The millennium development goals report 2005*; United Nations Dept. of Public Information: New York, 2005; pp 43.
21. Bellah, E. R. N.; Bellah, R. N.; Tipton, S. M.; Sullivan, W. M.; Madsen, R.; Swidler, A. *Habits of the heart: Individualism and commitment in American life*; University of California Press: Oakland, CA, 2007.
22. Colby, A.; Ehrlich, T.; Beaumont, E.; Rosner, J.; Stephens, J., Higher Education and the Development of Civic Responsibility. In *Civic Responsibility and Higher Education*, Ehrlich, T., Ed.; The American Council on Education and Oryx Press: Westport, CT, 2000; pp xii−xliii.
23. Stolle, D.; Hooghe, M. *Br. J. Polit. Sci.* **2005**, *35* (01), 149–167.
24. Youniss, J.; Bales, S.; Christmas-Best, V.; Diversi, M.; McLaughlin, M.; Silbereisen, R. *J. Res. Adolesc.* **2002**, *12* (1), 121–148.
25. Sax, L. Citizenship Development and the American College Student. In *Civic Responsibility and Higher Education*; Ehrlich, T., Ed.; American Council on Education and Oryx Press: Westport, CT, 2000; pp 3−17.
26. Zlotkowski, E. The case for Service Learning. In *Higher Education and Civic Engagement: International Perspectives*; Crowther, D., McIlrath, L., Labhrainn, L. M., Eds.; Ashgate Publishing Company: Burlington, VT, 2007.
27. Abbott, J.; Ryan, T. *Educ. Can.* **1999**, *39*, 8–11.
28. Schroeder, C. C. New students—New learning styles. *Change: The Magazine of Higher Learning* **1993**, *25* (5), 21–26.
29. Palmer, P. J. Community, conflict, and ways of knowing: Ways to deepen our educational agenda. *Change: The Magazine of Higher Learning* **1987**, *19* (5), 20–25.
30. McGrath, M. R. *Fostering Civic Engagements in America's Communities*; Jossey-Bass: Hoboken, NJ, 2001; Vol. 89.

31. Checkoway, B. Renewing the civic mission of the American research university. *J. High. Educ.* **2001**, 125–147.
32. Harkavy, I. The role of universities in advancing citizenship and social justice in the 21st century. *Education, Citizenship and Social Justice* **2006**, *1* (1), 5–37.
33. Moon, J. D. Civic Education, Liberal Education and Democracy. In *Higher Education and the Practice of Democratic Politics*; Murchlan, B., Ed.; Kettering Foundation: Dayton, OH, 1991; pp 196–207.
34. Cassidy, S. Learning styles: An overview of theories, models, and measures. *Educ. Psychol.* **2004**, *24* (4), 419–444.
35. Abott, J.; Marchese, T. *AAHE Bull.* **1996**, *48*, 3–6.
36. Sternberg, R. J.; Grigorenko, E. L. *Teaching for successful intelligence: To increase student learning and achievement*; Corwin Press: Thousand Oaks, CA, 2007.
37. Sternberg, R. *Successful Intelligence: How Practical And Creative Intelligence Determine Success In Life;* Penguin Putnam Inc.: New York, 1996.
38. Jacoby, B. *Building Partnerships for Service-Learning*; Jossey-Bass: San Fransisco, CA, 2003.
39. DeBoer, G. E. Scientific Literacy: Another Look at its Historical and Contemporary Meanings and Its Relationship to Science Education Reform. *J. Res. Sci. Teach.* **2000**, *37* (6), 582–601.
40. National Education Association of the United States. Commission on the Reorganization of Secondary Education. *Cardinal principles of secondary education. A report of the Commission on the reorganization of secondary education*; Govt. print. off.: Washington, DC. 1928.
41. Boyer, E. L. *Scholarship reconsidered: priorities of the professoriate*; 093105043X; Carnegie Foundation for the Advancement of Teaching: Princeton, N.J., 1990; pp 15–25.
42. Freire, P. *Pedagogy of the oppressed*; Herder and Herder: New York, 1970.
43. Freire, P.; Freire, A. M. A. J. *Pedagogy of hope : reliving Pedagogy of the oppressed*; Bloomsbury Academic: London, 2014.
44. Boyer, E. L. The scholarship of engagement. *Bull. Am. Acad. Arts Sci.* **1996**, 18–33.
45. Harkavy, I. *Education, citizenship and Social Justice* **2006**, *1* (1), 5–37.
46. Treagust, D. F. General instructional methods and strategies. In *Handbook of research on science education*; Abell, S. K., Lederman, N. G., Eds.; Routledge, Taylor and Francis Group: New York, 2007; Vol. 1, pp 373–391.
47. Domin, D. S. *J. Chem. Educ.* **1999**, *76* (4), 543.
48. McComas, W. F. 21st Century Skills. In *The Language of Science: An Expanded Glossary of Key Terms and Concepts in Science Teaching ad Learning*; McComas, W. F., Ed.; SensePublishers: Rotterdam, The Netherlands, 2014; pp 1–2.
49. Hofstein, A.; Lunetta, V. N. The laboratory in science education: Foundations for the twenty-first century. *Sci. Educ.* **2004**, *88* (1), 28–54.
50. Lunetta, V. N.; Hofstein, A.; Clough, M. P. Learning and teaching in the school science laboratory: An analysis of research, theory, and practice. In

Handbook of research on science education; Abell, S. K., Lederman, N. G., Eds.; Routledge, Taylor and Francis Group: New York, 2007; Vol. 1, pp 393−441.

51. Woods, D. R. *Ind. Eng. Chem. Res.* **2014**, *56*, 5337–5454.
52. Bell, S. *The Clearing House* **2010**, *83* (2), 39–43.
53. Blumenfeld, P. C.; Soloway, E.; Marx, R. W.; Krajcik, J. S.; Guzdial, M.; Palincsar, A. *Educ. Psych.* **1991**, *26* (3-4), 369–398.
54. Thomas J. W. *A Review of Research on Project Based Learning*; 2008; downloaded from http://www.newtechnetwork.org/sites/default/files/news/pbl_research2.pdf (accessed Aug 05 2014).
55. Larmer, J. *Project-Based Learning vs. Problem-Based Learning vs. X-BL. EDUTOPIA*; available at http://www.edutopia.org/blog/pbl-vs-pbl-vs-xbl-john-larmer (accessed August 05, 2014).
56. Sandi-Urena, S.; Cooper, M.; Stevens, R. *J. Chem. Ed.* **2012**, *89* (6), 700–706.
57. Sandi-Urena, S.; Cooper, M. M.; Gatlin, T. A. *Chem. Educ. Res. Pract.* **2011**, *12* (4), 434–442.
58. Kiefer, A. M.; Bucholtz, K. M.; Goode, D. R.; Hugdahl, J. D.; Trogden, B. G. *J. Chem. Educ.* **2012**, *89* (5), 685–686.
59. MacDonald, G. *J. Chem. Educ.* **2008**, *85* (9), 1250.
60. Johnson, J. R.; Savas, C. J.; Kartje, Z. a.; Hoops, G. C. *J. Chem. Educ.* **2014**, *9* (7), 1077–1080.
61. Millard, J. T.; Chuang, E.; Lucas, J. S.; Nagy, E. E.; Davis, G. T. *J. Chem. Educ.* **2013**, *90* (11), 1518–1521.
62. de los Santos, D. M.; Montes, A.; Sánchez-Coronilla, A.; Navas, J. S. *J. Chem. Educ.* **2014**, *9* (9), 1481–1485.
63. Tsaparlis, G.; Gorezi, M. *J. Chem. Educ.* **2007**, *84* (4), 668.
64. Robinson, J. K. *Anal. Bioanal. Chem.* **2013**, 1–7.
65. Palkendo, J. A.; Kovach, J.; Betts, T. A. *J. Chem. Educ.* **2013**, *91* (4), 579–582.
66. Wells, G.; Haaf, M. *J. Chem. Educ.* **2013**, *90* (12), 1616–1621.
67. Sadik, O. A.; Noah, N. M.; Okello, V. A.; Sun, Z. *J. Chem. Educ.* **2013**, *91* (2), 269–273.
68. Juhl, L.; Yearsley, K.; Silva, A. J. *J. Chem. Educ.* **1997**, *74* (12), 1431.
69. Kalivas, J. H. *J. Chem. Educ.* **2008**, *85* (10), 1410.
70. Anunson, P. N.; Winkler, G. R.; Winkler, J. R.; Parkinson, B. A.; Schuttlefield Christus, J. D. *J. Chem. Educ.* **2013**, *90* (10), 1333–1340.
71. Draper, A. J. *J. Chem. Educ.* **2004**, *81* (2), 221.
72. Wenzel, T. J. In *Collaborative and Project-Based Learning in Analytical Chemistry*; ACS Symposium Series 970; American Chemical Society: Washington, DC, 2007; pp 54−68.
73. Guo, Y.; Young, K. J.; Yan, E. C. Y. Guided Inquiry and Project-Based Learning in Biophysical Spectroscopy. In *Teaching Bioanalytical Chemistry*; Hou, H. J. M., Ed.; American Chemical Society: Washington, DC, 2013; Vol. 1137, pp 261−291.
74. Hopkins, T. A.; Samide, M. *J. Chem. Educ.* **2013**, *90* (9), 1162–1166.

75. Bringle, R. G.; Hatcher, J. A. *Michigan J. Comm. Serv. Learn.* **1995**, *2* (1), 112–122.
76. Bringle, R. G.; Hatcher, J. A. *J. Higher Educ.* **2000**, 273–290.
77. Saitta, E.; Bowdon, M.; Geiger, C. *J. Sci. Educ. Technol.* **2011**, *20* (6), 790–795.
78. Kammler, D. C.; Truong, T. M.; VanNess, G.; McGowin, A. E. *J. Chem. Educ.* **2012**, *89* (11), 1384–1389.
79. Dewey, J. *Education and Experience*; Kappa Delta Pi: Indianapolis, IN, 1998.
80. Giles, D. E.; Eyler, J. *Michigan J. Comm. Serv. Learn.* **1994**, *1* (1), 77–85.
81. Dewey, J. *Democracy and Education: An Introduction to the Philosophy of Education*; The Macmillan Company: New York, 1916; pp xii, 11, 434.
82. Rocheleau, J. *Theoretical roots of service- learning: Progressive education and the development of citizenship*. In *Service-Learning: History, Theory, and Issues*; Speck, B. W., Hoppe, S. L., Eds.; Prager Publishers: Westport, CT, 2004.
83. Hepburn, M. A. *Theory into Practice* **1997**, *36* (3), 136–142.
84. Speck, B. W.; Hoppe, S. L. *Service-learning: History, theory, and issues*. In *Service-learning: History, theory, and issues*; Speck, B. W.;, Hoppe, S. L., Eds.; Prager Publishers: Westport, CT, 2004.
85. Astin, A. W. *Higher education and the concept of community*; University of Illinois at Urbana-Champaign: Champaign, IL, 1993.
86. *Life, learning, and community: Concepts and models for service-learning in biology*; Brubaker, D. C., Ostroff, J. H., Eds.; American Association for Higher Education: Washington, DC, 2002; p 174.
87. Kennell, J. C. Educational benefits associated with service-learning projects in biology curricula. In *Life, Learning, and Community: Concepts and Models for Service-Learning in Biology*; Brubaker, D. C., Ostroff, J. H., Eds.; American Association for Higher Education: Washington, DC, 2002; p 174.
88. Sutheimer, S. *J. Chem. Educ.* **2008**, *85* (2), 231.
89. Abel, C. F. A Justification of the philanthropic model. In *Service-Learning: History, Theory, and Issues*; Speck, B. W., Hoppe, S. L., Eds.; Prager Publishers: Westport, CT, 2004.
90. Sementelli, A. A critique of the philanthropic model. In *Service-Learning: History, Theory, and Issues*; Speck, B. W., Sherry, L. H., Eds.; Greenwood Publishing Group: Westport, CT, 2004.
91. Furco, A.; Billig, S. *Service-learning: the essence of the pedagogy*; Information Age Pub.: Greenwich, CT, 2002; p x, 286.
92. Hollander, E. L.; Saltmarsh, J. *Academe* **2000**, *86* (4), 29–32.
93. Bennett, W. J.; Nunn, S. *A Nation of Spectators: How Civic Disengagement Weakens America and What We Can Do about It*; University of Maryland: College Park, MD, 1998.
94. Codispoti, F. *A justification of the communitarian model*. In *Service-Learning: History, Theory, and Issues*; Speck, B. W., Hoppe, S. L., Eds.; Prager Publishers: Westport, CT, 2004.
95. Hatcher-Skeers, M.; Aragon, E. *J. Chem. Educ.* **2002**, *79* (4), 462.

96. Samide, M.; Akinbo, O. *Anal. Bioanal. Chem.* **2008**, *392* (1), 1–8.
97. Akinbo, O. T. Teaching Bioanalytical Chemistry in an Undergraduate Curriculum: The Butler University Analytical Chemistry Curriculum as an Example Model. In *Teaching Bioanalytical Chemistry*; Hou, H. J. M., Ed.; American Chemical Society: Washington, DC, 2013; Vol. 1133, pp 23−56
98. Vivelo, E; Mayfield, M.; Han. S. *Determination of Trace Elements and Minerals Present in Drinking Water. A report submitted in partial fulfillment of requirements for Analytical Chemistry 1*; Department of Chemistry: Butler University,
99. Ikem, A.; Odueyungbo, S.; Egiebor, N. O.; Nyavor, K. *Sci. Total Environ.* **2002**, *285* (1), 165–175.
100. Yekdeli Kermanshahi, K.; Tabaraki, R.; Karimi, H.; Nikorazm, M.; Abbasi, S. *Food Chem.* **2010**, *120* (4), 1218–1223.
101. Plotan, M.; Frizzell, C.; Robinson, V.; Elliott, C. T.; Connolly, L. *Food Chem.* **2013**, *136* (3), 1590–1596.
102. National Research Council. *National Science Education Standards*; The National Academies Press: Washington, DC, 1996.
103. Bell, S. *The Clearing House* **2010**, *83* (2), 39–43.
104. Larmer, J.; Mergendoller, J. R. *8 Essentials for Project-Based Learning*, downloaded from Buck Institute of Education through http://bie.org/object/document/8_essentials_for_project_based_learning (accessed, Aug 05, 2014).
105. Laursen, S.; Hunter, A.-B.; Seymour, E.; Thiry, H.; Melton, G. *Undergraduate research in the sciences: Engaging students in real science*; John Wiley & Sons: New York, 2010.
106. Seymour, E.; Hunter, A. B.; Laursen, S. L.; DeAntoni, T. *Sci. Educ.* **2004**, *88* (4), 493–534.
107. Lopatto, D.; Tobias, S. *Science in solution: The impact of undergraduate research on student learning*; Council on Undergraduate Research: Washington, DC, 2010.
108. Lopatto, D. *Peer Review* **2006**, *8* (1), 22−25; Lopatto, D. *Sci. Educ.* **2007**, *6* (4), 297−306.
109. Shaffer, C. D.; Alvarez, C. J.; Bednarski, A. E.; Dunbar, D.; Goodman, A. L.; Reinke, C.; Rosenwald, A. G.; Wolyniak, M. J.; Bailey, C.; Barnard, D. *Sci. Educ.* **2014**, *13* (1), 111–130.
110. National Research Council (U.S.) Committee on Undergraduate Biology Education to Prepare Research Scientists for the 21st Century. *BIO2010: Transforming Undergraduate Education for Future Research Biologists*; 2003.
111. Brewer, C. A.; Smith, D. *Vision and Change in Undergraduate Biology Education: A Call to Action*; American Association for the Advancement of Science: Washington, DC, 2011.

Chapter 11

Instrumental Analysis at Seattle University: Incorporating Environmental Chemistry and Service Learning into an Upper-Division Laboratory Course

Douglas E. Latch*

Department of Chemistry, Seattle University, Seattle, Washington 98122
*E-mail: latchd@seattleu.edu.

This chapter describes efforts to revitalize Seattle University's advanced analytical chemistry course, Instrumental Analysis, by incorporating research-like laboratory experiments that focused on local environmental issues via service-learning projects. These projects focused on measuring concentrations of several pollutants in nearby water, sediment, and soil that are of interest to a local community group. Specifically, students were involved in collecting samples and developing analytical methods for the quantification of lead, polychlorinated biphenyls, and pyrethroid pesticides using modern scientific instrumentation. Pre- and post-course surveys were administered to determine how the students' attitudes changed after participating in laboratory based service-learning projects and readings related to the service-learning projects. Overall, the students self-reported modest gains in their interest in and understanding of analytical and environmental chemistry. Short answer post-course survey questions revealed that most students had favorable impressions of the environmentally relevant research undertaken as their service-learning projects.

© 2014 American Chemical Society

Introduction

Over the last two decades, many calls have been made to reform science, technology, engineering, and mathematics (STEM) education. One of the outcomes of these efforts has been the emergence of many STEM courses that have incorporated service-learning components (*1–24*). In addition to providing valuable real-world context to the students enrolled in these classes, these service-learning courses are designed to provide benefits to the community. In chemistry departments, such service-learning courses have been designed to benefit local communities in several ways, most notably by providing the public with data about local environmental conditions (*4–11*) and by contributing to the general education and scientific literacy of the populace (*5–22*).

Several accounts of environmentally themed service-learning projects now exist in science education websites, journals, and books (*3–11, 23, 24*). The main purpose of this chapter is simply to add to that growing body of literature by describing how an upper division analytical chemistry course at Seattle University was modified to include environmental service-learning projects. A secondary objective is to provide feedback from students regarding their experiences during the course.

Description of the Course

Instrumental Analysis at Seattle University is an upper-level combined lecture and laboratory course that meets twice weekly throughout a twelve week quarter. Each class session is scheduled for approximately 4.5 hours, with roughly the first hour devoted to lecture and the rest of the time devoted to the laboratory. It is designed to introduce students to the fundamental concepts and common laboratory techniques used in chemical analyses involving instrumentation. The course is the second portion of our two-quarter analytical chemistry sequence and is required for all B. S. Chemistry majors (and is an elective for our B. S. Biochemistry and B. A. Chemistry majors, as well as for the Chemistry minor). A few students enroll in Instrumental Analysis in the quarter immediately after taking the introductory analytical chemistry prerequisite course. Most students, however, will have at least one quarter (and as many as four quarters) separation between enrollments in the two analytical chemistry courses. In the introductory analytical course, students individually work on projects aimed at perfecting their basic laboratory skills and treatment of data. In Instrumental Analysis, the students work together in small groups.

Historically, Instrumental Analysis students worked in teams of two or three on several (six to eight) "cookbook" laboratory experiments. These cookbook labs are relatively straightforward experiments in which students follow written protocols to achieve results that are well known to the instructor. Recently, we have been developing the course to include more civic-minded projects dealing with important environmental and ecological issues. In addition to providing tangible real-world context to the laboratory portion of the course, these new projects have been designed to give the students an authentic research-like experience.

An excerpt of the course syllabus showing the personal learning goals and academic learning outcomes outlined for the students is given below:

Personal Learning Goals

My hope and expectation is that by participating actively and earnestly in class and in the laboratory you will:

i. Work in a comfortable, confident, and conscientious manner in the laboratory.
ii. Work effectively and supportively with your colleagues.
iii. Identify and implement study practices that work best for you.
iv. Enjoy the subject and look forward to coming to class.
v. Behave in a professional, courteous, and well-organized manner.
vi. Connect how your chemistry skills and knowledge can be impactful to the world around you.

Academic Learning Outcomes

In addition to the personal learning goals listed above, upon successful completion of this course, you will be able to:

i. Operate sophisticated scientific instruments commonly found in chemistry laboratories.
ii. Develop analytical methods for detecting pollutants in various matrices.
iii. Choose and successfully employ appropriate instrumental and calibration techniques depending on the particular experimental parameters.
iv. Describe the underlying principles involved in spectroscopy, chromatography, and mass spectrometry.
v. Collect, critique, and use spectroscopic, chromatographic, and mass spectrometric data to determine the identity of unknown analytes and to quantify their concentrations.
vi. Formulate and disseminate sound scientific ideas through written, visual, and oral communication.

Several of these goals (particularly *vi*) and objectives (especially *ii*, *v*, and *vi*) were written with the students' service-learning projects in mind. In this laboratory-heavy course, students are split into groups and rotate through several experiments during the quarter. The experiments have been designed to allow for students to gain hands-on experience operating many instruments encountered in modern chemistry laboratories; such instruments typically include absorption and fluorescence spectrophotometers, an atomic absorption spectrophotometer (AAS), a Fourier-transform infrared (FTIR) spectrophotometer with an attenuated total reflectance (ATR) attachment, a high performance liquid chromatograph (HPLC), a gas chromatograph-mass spectrometer (GC-MS), and liquid chromatograph-tandem mass spectrometer (LC-MS-MS).

Introduction to the Service-Learning Projects

Our community partner was the local citizens' group *People for Puget Sound*. I connected with this group through a Seattle University colleague (Biology Prof. Lindsay Whitlow) who had an ongoing relationship with them through his ecology course (*3*). *People for Puget Sound* is an organization founded in 1991 to "protect and restore the health of our land and waters through education and action" (*25*). They are active in educating students and citizens about local ecological and environmental issues, leading efforts to restore areas heavily impacted by anthropogenic influences, and lobbying policymakers. To effectively carry out these efforts, they must have an understanding of the extent to which different sites have been contaminated by various pollutants. To this end, we looked to partner with *People for Puget Sound* to test soil and water samples in the region for levels of several contaminants of concern. In order to identify particular projects that would be both of interest to *People for Puget Sound* and suitable for the Instrumental Analysis students to tackle, I met with a representative from *People for Puget Sound*. Because the Puget Sound region is quite urban and heavily impacted by industry, the *People for Puget Sound* have many contaminants that they are interested in tracking. We devised three projects that were of interest to *People for Puget Sound* and seemed feasible, though likely quite challenging, for advanced undergraduate chemistry majors equipped with modern instrumentation. Specifically, the Instrumental Analysis students aimed to:

i. Determine the lead content in soil samples taken from nearby Vashon Island and sites near the Seattle University campus. This project was designed to assess the extent of contamination caused by the Asarco smelter plume emanating from Tacoma;

ii. Measure concentrations of polychlorinated biphenyls (PCBs) in Duwamish River sediment samples to assess legacy contamination due to shipping and industry; and

iii. Measure levels of pyrethroid pesticides entering the Duwamish River estuary through urban runoff.

These three new projects each address local environmental issues and, importantly for the Instrumental Analysis course, rely on different instrumentation. The lead analysis was carried out using an atomic absorption spectrophotometer, the PCBs analysis was performed on a GC-MS, and the pyrethroid pesticides were analyzed by LC-MS-MS. That the three service-learning projects rely on three different instruments was critical in achieving one of the aims of the course: that each student would gain experience using several instruments found in modern chemistry laboratories. The lead analysis project was undertaken because of worries about historic pollution carried by the exhaust plume from the Asarco Smelter in Tacoma (*26*, *27*). In addition, Seattle University is an urban campus located near several major roadways that may have imparted surrounding areas with historic burdens of lead from leaded gasoline. Students used a modified version of an EPA method (*28*) to process soil samples that they collected from sites near campus and from other sites in the direct path of the Asarco Smelter plume. Groups who participated in this project at the beginning of the quarter focused on developing the procedure and testing its sensitivity and precision. Later groups used the developed procedure to quantify soil lead content while also running blanks and spike recovery experiments to identify potential matrix effects. The students used a flame AAS (Perkin Elmer Model AAnalyst 200) for their analyses and were able to detect lead levels as low as a few ppm w/w in soils (when accounting for the work up associated with EPA Method 3050B). Calibration curves were generated from a stock solution prepared by the Department of Chemistry's Laboratory Supervisor.

The Duwamish waterway in south Seattle is a Superfund site that has a legacy of industrial pollution (*29, 30*). We aimed to quantify one class of pollutant, PCBs, in Duwamish sediment using GC-MS (Agilent 6890 GC with autosampler and 5972 MS). Student groups attempted to develop a GC-MS method to separate and quantify PCBs in a certified reference sample. Unfortunately, this proved difficult and we were not able to develop a suitable method for the PCBs. Several groups (with my assistance) tried to develop a method, but we were unable to achieve appropriate sensitivity to detect the PCBs. Later groups developed methods for separating polycyclic aromatic hydrocarbons (PAHs), another important group of legacy contaminants at industrial sites. Even though students were unable to accomplish the goal of quantifying PCBs in Duwamish sediment, they still gained valuable first-hand experience operating a GC-MS, working through the stages of method development, and realizing the difficulties that are often encountered when developing a research project.

The third service-learning project was a multi-class collaborative effort to quantify pyrethroid pesticides in Duwamish water. Pyrethroids are emerging pollutants that are often present in urban runoff. The Instrumental Analysis students' role in this project was to use an LC-MS-MS method to quantify pyrethroid concentrations in samples that were collected by students in an ecology course and extracted by organic chemistry students. The students used an Agilent 1200 Series HPLC-triple quadrupole 6410 mass spectrometer system and isotopically enriched permethrin standards (*cis*-permethrin-$^{13}C_6$ and *trans*-permethrin-$^{13}C_6$ from Cambridge Isotope Labs, Inc.) to quantify pyrethroids and assess the performance of the analytical method. Stock solutions of several pyrethroids were prepared from neat commercial sources by a research student working with the course instructor. Calibration curves generated by the students from these stock solutions showed that the pyrethroids were detectable at low ng/L levels in the sampled water once the concentrating effect of the solid phase extraction work up was taken into account. Although detection limits and recoveries of the isotopically enriched standards were quite good, pyrethroids were not detected in any of the water samples analyzed. This project is described in greater detail elsewhere (*31*).

It is worth noting that the students rotate which experiment they are working on from week to week. This is a necessity given the availability of our advanced instrumentation and the longstanding departmental goal of having all student groups operate and become comfortable using each of the instruments during the quarter. An alternative way to structure environmental service-learning projects in an advanced laboratory course would be to assign each group of students to a particular project to work on throughout the quarter. In this case, students would be able to delve deeply into their projects and have more time to truly master the techniques and instrumentation used in their analyses. Rather than opt for this type of deep involvement in a single project, we opted for the breadth associated with a rotation of experiments. Thus, student groups only have two lab periods to work on any one experiment. As such, the students had to build upon what prior groups accomplished on these projects. For example, groups working on the lead project at the beginning of the quarter focused on developing and testing the analytical procedure and communicated their findings to subsequent groups. These later groups were then able to analyze samples that they had collected from the field. Unfortunately, this meant that a couple groups never actually collected and analyzed field samples while contributing to the overall projects. This is a drawback that concerned students (see Summary of Student Feedback section below), but was difficult to address at the outset of these projects.

Several steps were taken to prepare the students for the service-learning aspects of the course. To provide the students with a broad overview of the projects they were going to undertake throughout the quarter, at the beginning of the quarter I gave an approximately hour-long presentation describing the service-learning projects. This presentation touched on the general benefits of service-learning, gave a brief background on the *People for Puget Sound*, and provided specific details about the three service-learning projects. Details included the local environmental significance of the pollutants to be measured and the manner in which we would be approaching the project (e.g., by collaborating on projects, by developing new analytical procedures for samples with truly unknown concentrations of pollutants, and by making incremental improvements to the procedures from week to week and group to group). I also added environmentally themed primary literature assignments to the classroom portion of the course. In these class sessions, students worked in groups, with guided questions, reading and analyzing three analytical environmental articles relevant to the broad objectives of the course and which would be insightful to the service-learning projects.

As described elsewhere (*31*), an evening meeting was held for students enrolled in Seattle University's Instrumental Analysis, Organic Chemistry Laboratory, and Ecology courses to view the PBS Frontline documentary Poisoned Waters about the impacts of heavy anthropogenic pollution in the Puget Sound and Chesapeake Bay (*32*). The Duwamish River, which was the focus of two of the service-learning projects, drains into the Puget Sound, making this documentary particularly meaningful for the students. A mid-quarter visit to the field site was also scheduled where the students served on a restoration project run by Duwamish Alive, a non-profit citizen's group focused on protecting and restoring local habitats. During the quarter, the *People for Puget Sound* representative that I had previously met with visited the class and gave a guest lecture on environmental issues in the region. She touched on the projects that the students were working on as well as provided a broad overview of how anthropogenic chemicals enter the environment and affect ecosystem and human health. A final meeting between Seattle University's Instrumental Analysis and

Ecology students was held as a festive pizza party during the last week of labs to allow them to present and share with each other their results and experiences working on research-based projects of interest to the local community. The representative from *People for Puget Sound* who visited our class at the beginning of the quarter was also invited to attend this event, but she was unable to attend.

Because the environmental issues that *People for Puget Sound* wanted to address were new to the course, cookbook lab instructions were not available. In addition, because the samples came from real-world sites with unknown levels of contaminants, we did not know what results to expect from the projects. These factors caused several challenges that added to the complexity of the course and initially increased students' anxiety levels about what they needed to complete to achieve satisfactory grades. The anxiety regarding grades arose somewhat from the very results-focused grading policy that the students were accustomed to in the pre-requisite Quantitative Analysis course. The students' anxiety was largely alleviated by clearly articulating to them that their laboratory grades in Instrumental Analysis were not to be based on how close they get to an accepted value for their unknowns (as they were in Quantitative Analysis), but rather on their effort in developing and troubleshooting analytical procedures, working up and presenting their acquired data as tables and figures, and the quality in which they write an experimental section describing the work they did in the laboratory. It should also be noted that there indeed were no accepted values for the Instrumental Analysis environmental projects, because the concentrations of target substrates were truly unknown (to both the students and the instructor). After completing their first post-laboratory assignment and seeing how it was graded, the students were much less stressed about how their laboratory assignments were to be assessed. An added feature of this style of laboratory experiment is that the students more closely modeled the collaborative nature in which professional analytical/environmental chemists approach their work. Students working on a given project early in the quarter focused on preparing robust analytical methods. Later groups tested the methods developed by prior groups. The final groups used the developed and vetted methods to actually measure contaminant concentrations. Because of the division of labor and collaboration across groups, we anticipated that the methods the students developed and the data they acquired would improve as the quarter proceeded. Importantly, dividing the tasks in this way, rather than having a project assigned to each group to work on throughout the quarter, met one of the academic objectives of the course by allowing each group to gain experience using all of the instrumentation in the laboratory. By working on environmentally relevant projects in collaboration with their peers and an interested community partner and in the manner that professional laboratories operate, it was anticipated that two additional tangential benefits would be that students would develop a deeper sense of belonging to the chemistry discipline and more fully appreciate the role of chemists in society.

All students participated in the service-learning projects. They were required to complete post-lab assignments after each lab experiment (service-learning and cookbook). The points allotted for the service-learning lab assignments amounted to eighteen percent of the total points for the quarter (the cookbook labs were worth the same number of points). An additional eight percent of the total points for the course were allotted for participation in service-learning and in-class activities and for completing pre- and post-course surveys relating to their service-learning projects.

Summary of Student Feedback

Students were surveyed before and after completing their service-learning projects. The surveys were administered via the course's Angel website. Students were given a window of several days to complete each of the surveys. Each survey consisted of fifteen Likert-style questions. Six additional questions were added to the post-course survey to allow students to respond in more detail. Each survey was designed so that the students could complete them in less than thirty minutes.

Because the course is typically offered only once per year and all students were required to participate in the service-learning laboratory projects, a control group was unavailable. Thus, analysis was limited to the pre- and post-course surveys. I also note that given the very small sample size (i.e., one class of fourteen students), caution should be taken to not place too much weight on the survey results. Rather than using this feedback to rigorously assess the effectiveness of the course, these should simply be used as a way to probe students' general satisfaction with the environmental service-learning aspects of the course and to use their comments in making improvements to this course and others like it.

Numeric Pre- and Post-Course Surveys

Results of the pre- and post-course surveys are shown in Figure 1. In general, students self-reported modest gains in their background, understanding, and interest in analytical and environmental chemistry. These two areas of chemistry are highly relevant to both the course material and the service-learning project. Students also reported that their sense of "feel[ing] like a 'real' chemist or biochemist" improved during the quarter, as did their level of comfort using instrumentation.

Qualitative Student Feedback

The students also provided valuable written feedback about the service-learning portion of the course. The following section provides a few representative samples of students' responses to end-of-course survey questions. I note that all fourteen students completed the survey and each gave thoughtful responses to nearly every question and that the summary of responses given below were selected from those given by all of the students in the class without knowledge of which student gave the response (i.e., responses are not necessarily from the same students from question to question). The students overwhelmingly indicated that participation in the service-learning projects increased their motivation and interest levels in the course (see responses to post-course survey Question 1 in Figure 2).

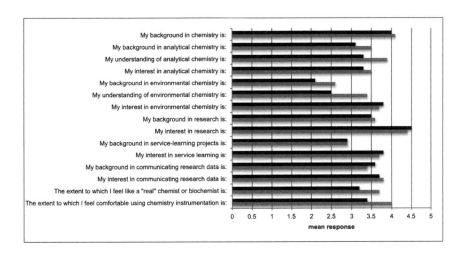

Figure 1. Results of pre- and post-course surveys (1 = weak, 5 = strong; top bar = pre-course responses, bottom bar = post-course responses). Students self-reported modest gains in several areas.

Question 1. Did the service-learning portion of the course, in which you developed methods and analyzed local soil and water samples, help you learn the material or help to motivate you to do well on these projects? If so, explain how.

- Definitely. It was very different thinking I was actually helping a real cause. I took more time with my samples and I did more careful work. I took it much more seriously knowing that I was doing legitimate science and now just pumping out data for a lab assignment.
- Yes, the real world application of the experiments was highly motivating. I felt more anticipation for finding the results because I knew that they would be impacting real people and not just my grade.
- The service-learning component definitely made me more interested in environmental chemistry, and I suppose it also motivated me to do well on these projects, since the results actually meant something in that it had real life applications... The project also helped me understand AAS [atomic absorption spectroscopy] more, and how the instrument works.
- Yes, I found the project very engaging. It was nice to be providing data to a larger organization which serves the community. The project increased my familiarity with real practices in environmental and analytical chemistry and made the course more interesting.

Figure 2. Post-course survey question regarding student learning and motivation along with some representative student responses.

The presentation by the representative from *People for Puget Sound* was generally well received, though some would have preferred to hear from a scientist rather than from somebody in the lobbying portion of the organization (see responses to post-course survey Question 2 in Figure 3).

> **Question 2.** To what extent did the presentation from the representative for the *People for Puget Sound* help to place your work into a real-world context? Please provide your general feeling about her seminar and how it contributed to/hindered your learning.
>
> - I thought the presentation dealing with the People for Puget Sound placed our work into a real-world context a great amount. I thought her seminar helped me learn more about how these contaminants are entering the environment and how they can have a negative effect on animals and humans... Overall, I personally thought the presentation was very helpful and interesting because it allowed us to see where these contaminants are originating from and what the People for PS are doing in attempt to prevent it or fix it.
> - I really valued the opportunity to hear a community organizer come into the classroom and speak to the urgent issues with which my generation will be challenged. Maybe I am making assumptions, but I often feel a strong sense of apathy from my peers and I know personally I need to be taking more action around the issues she was presenting on. I wish that there were closer ties between community organizations and the university. In my opinion, the university should be a resource for the community, not only for the education of students, but also to generate ideas/research that genuinely helps the larger community.
> - While it was clear that the representative was not a scientist, it was still interesting to listen to the ideas presented.
> - I liked learning about environmental problems that are relevant to our local area. What she talked about in her presentation was mostly new to me, as I haven't had really and prior education about environmental science. It really helped put what we were doing and learning in the lab into perspective.

Figure 3. Post-course survey question regarding the community partner's in-class presentation along with some representative student responses.

Students had favorable impressions of the scientific literature examples that were analyzed as a class to help prepare them for their service-learning projects and provide additional context to the environmental aspects of the course (see responses to post-course survey Question 3 in Figure 4).

> **Question 3.** To what extent did reading scientific papers on analytical environmental chemistry help to place your work into a real-world context? Please provide your general feeling about how using the scientific literature contributed to/hindered your learning.
>
> - It was helpful to get a big picture idea from many environmental chem ideas. It was also helpful to understand quality control/assurance.
> - I thought the scientific papers helped place my work into a real-world context by a great deal. By reading the articles and seeing how other chemists are analyzing samples from the environment to determine the extent to which they are contaminants allowed me to know more about their process. In addition, seeing how the different analytical techniques are applied in these papers gave us a sense of what chemicals can be measured with (or preferred with) which instrument.
> - Reading the papers on analytical environmental chemistry helped me see how the techniques we're learning about in class are applied to real-world research.
> - I think it was a good addition to the course. Going over paper helped us understand analytical chemistry articles and the terminology among other things. The scientific papers helped me put into context the things we were learning to do in the lab, and what and how they can be applied in the real world.

Figure 4. Post-course survey question regarding the use of environmentally themed scientific literature activities during the lecture portion of the course along with some representative student responses.

For many students, the service projects illuminated or reinforced how chemists can positively impact the community (see responses to post-course survey Question 4 in Figure 5).

> **Question 4.** Did the service-learning portion of the course help you to see how chemists can make positive contributions to the community? If so, what can be done to this course to further strengthen that feeling?
>
> - Yes. It was nice to see the ways a chemist may pursue a career that is helpful to society/environment.
> - Yes, I definitely think it does. Being able to analyze these samples and determine the certain components that are present can lead us to fix them and eliminate the possible contaminants in the environment. I think having a small service-learning project assigned to each group at the beginning of the quarter that they can work on throughout the quarter could enhance this feeling.
> - Yes, in the past I have struggled to connect my interests in environmental justice and chemistry. This course made it clear that both can support and push each other forward.
> - Since Organic Chem, I really wanted a real life application and this class was outstanding in incorporating outside research into the class room... I think we should make all the labs based on service learning portion. I mean not all of them have to be "big scale" projects but any real life application works.

Figure 5. Post-course survey question regarding the students' connection to the community along with some representative student responses.

Most students were happy with the overall mix of traditional, straightforward cookbook-style laboratory experiments and the more open-ended service-learning projects (see responses to post-course survey Question 5 in Figure 6).

> **Question 5.** To what extent would you prefer or not prefer having all of the labs in this course be relatively simple "cookbook" labs that have final results that are known to the instructor? Please explain your answer.
>
> - It's a good balance I think. I think we still need some "cookbook" labs that we can test ourselves against and get a positive or negative affirmation of our abilities against the known result.
> - I think having some of the labs that were relatively simple "cookbook" labs, where the results are known, allows us to check how we are running our experiments and whether or not we are doing it right. Also, keeping some labs consistent allows us to monitor how we are doing and how the instruments are working.
> - I think the mix we had of "cookbook" and non "cookbook" labs was good. With the labs that didn't have specific instructions, it would get a bit frustrating because we often didn't know what we're doing, partly because our task was to make an SOP or the instrument was not functioning properly. The "cookbook" labs were kind of a break from the frustration, since we had set instructions and results to get through with calibration curves and equations. The mix is good because we won't always have instructions to follow, and it will be our responsibility to make the protocol or develop a method.
> - It is nice having cookbook labs, which are known to work to learn the basic technique. But having labs that aren't so simple is interesting and educational, I think. I liked it. Perhaps we could have cookbook labs, and then the technique we learn in those labs we could apply to service learning or more advanced untested labs experiments.

Figure 6. Post-course survey question regarding the mix of laboratory projects in the class along with some representative student responses.

The students also gave important suggestions for improving the course (see responses to post-course survey Question 6 in Figure 7). Some indicated that all students should be more directly involved in sample collecting so that they have a better sense of ownership and connection to the samples. They also expressed some frustration about the lack of information about what to do in lab and for assignments, especially when the instruments or experimental procedures were not working as expected.

Question 6. How would you improve the service-learning portion of the course?

- Provide more background from the beginning.
- I think maybe scheduling in a field trip to a site related to the service-learning project or visiting a governmental lab that is doing environmental chemistry would be really exciting.
- I don't know if this would be difficult to do, but I think it'd be cool if every group had a separate project to work on (i.e., AAS for lead determination in soil, LC-MS permethrin in water, etc.) that they can report to the class at the end of the quarter. The presentation would also include an analytical technique, just as we did this quarter. However, the problem with this is the time limit and that there are many instruments we need to use in this class.
- Allow groups to work on the projects throughout the quarter to give it continuity and allow the formation of a big picture understanding of the process.

Figure 7. Post-course survey question regarding students' ideas for improving the course along with some representative student responses.

Discussion and Summary

The students' responses to the survey questions overwhelmingly indicate that they enjoyed being involved in socially relevant research projects. Indeed, many of the students' written responses press for increased levels of involvement in this type of field research and dissemination of the data that they acquire. In addition to providing motivation to do well on their experiments, the service projects also challenged the students to use their science skills and learn new techniques and instrumentation. With such drastic changes made to this course, it is important to note that the students in this class performed at least as well in the lab as any of the prior Instrumental Analysis classes that I have taught at Seattle University. For those considering adding service learning or research-based assignments to their laboratory courses, I also note that my course evaluations (i.e., Student Perception of Teaching forms) remained high despite the uncertainty associated with introducing several new, open-ended laboratory assignments. They also met the challenge of working together to develop sound experimental protocols and measure trace-level pollutants that prior students were never asked to do.

From my own impression of working with them in the lab and from their written responses in their post-course survey, it appears that this group of students worked with a greater sense of purpose and appreciated the opportunity to provide data to the community. The results of the surveys are consistent with the thoughts, concerns, and comments that the students shared with me during laboratory sessions and office hours.

The students suggested several ways to improve the course, and I agree with their sentiments. The benefits of having all of the students visit the study sites and collect samples likely outweigh the logistical problems associated with arranging a field-sampling visit for the entire class. This will help all students to feel connected to the community in which they are collecting samples and makes them even more vested in the analysis of the samples. Now that we have developed laboratory protocols, future students will be able to analyze more samples with better reliability and confidence in their results. This will hopefully enable us to address another of the students' concerns: what happens to their results? Because of issues with the analytical methods, we could not in good faith disseminate the results of these service-learning projects. It may be possible to do so in subsequent years given the significant foundation put in place by this class. In the future, student generated data may be validated using a variety of techniques such as spike/recovery experiments, analysis of method blanks and certified reference materials, and assessment of the recovery of chromatographic method and injection standards. Future validated data could potentially be shared with the Washington State Department of Ecology, the *People for Puget Sound*, and/or other community organizations. Inviting the community partner and other interested community members to an end of quarter town hall style meeting to disseminate our findings is also desirable. In addition, care must be taken to streamline and make more uniform the instructions and assignments so that the students have a better idea of what they need to do each week, especially when an experiment does not work as planned.

Future iterations of this course will likely include additional environmentally themed projects that arise according to community need or to more broadly address gaps in current scientific knowledge within the environmental science community. These projects may replace either some of the current environmental projects or some of the more traditional cookbook experiments.

Overall, the academic service-learning projects were well received by the students and contributed to their learning and enjoyment of the course. It can be difficult to thoroughly and deeply engage students in chemistry laboratory courses, and these service-learning projects appear to have worked in that regard. The students also reported that the service-learning component of the course motivated them to work more carefully than in their traditional laboratory courses, created a stronger bond to the local community, and clearly illustrated the value of scientists in the public realm. Lastly, the students' academic performance and breadth of material covered in the class was not hindered by participating in the service-learning projects; the course was designed such that the students in this course participated in the same number of laboratory experiments and got hands-on experience with the same instrumentation as those in prior (non-service-learning) iterations of the course.

Acknowledgments

I thank Jeffrey Anderson and Seattle University's Academic Service-Learning program for providing valuable guidance, feedback, and support as I designed and implemented the service-learning aspects of the course. I gratefully acknowledge Dr. Lindsay Whitlow and Dr. Peter Alaimo for helpful discussions and their efforts in developing the pyrethroid pesticide portion of project. I also thank my Seattle University analytical chemistry colleague, Dr. Kristy Skogerboe, for providing insights into the laboratory experiments and for her support during the development of the environmental focus in the Instrumental Analysis course. Lastly, I thank the *People for Puget Sound* for their collegiality and support in developing the service-learning projects.

References

1. *Acting Locally: Concepts and Models for Service-Learning in Environmental Studies*; Ward, H., Ed.; American Association for Higher Education: Washington, DC, 1999.
2. Eyler, J.; Giles, D. E. Jr. *Where's the Learning in Service Learning?* Jossey-Bass: San Francisco, 1999.
3. Whitlow, W. L; Hoofnagle, S. Mud, Muck, and Service. Action Research on Direct and Indirect Service Learning in Ecology. *Science Education and Civic Engagement* **2011**, *3*, 57–65.
4. Kesner, L.; Eyring, E. M. Service–Learning General Chemistry: Lead Paint Analyses. *J. Chem. Educ.* **1999**, *76*, 921–923.
5. Draper, A. J. Integrating Project-Based Service-Learning into an Advanced Environmental Chemistry Course. *J. Chem. Educ.* **2004**, *81*, 221–224.
6. Landolt, R. G. Incorporating Chemical Information Instruction and Environmental Science into the First-Year Organic Chemistry Laboratory. *J. Chem. Educ.* **2006**, *83*, 334–335.
7. Gardella, J. A., Jr.; Milillo, J. A.; Sinha, T. M.; Oh, G.; Manns, D. C.; Coffey, E. Linking Community Service, Learning, and Environmental Analytical Chemistry. *Anal. Chem.* **2007**, *79*, 811–818.
8. Suthheimer, S. Strategies To Simplify Service-Learning Efforts in Chemistry. *J. Chem. Educ.* **2008**, *85*, 231–233.
9. Kalivas, J. H. A Service-Learning Project Based on a Research Supportive Curriculum Format in the General Chemistry Laboratory. *J. Chem. Educ.* **2008**, *85*, 1410–1415.
10. Donaghy, K. J.; Saxton, K. J. Service Learning Track in General Chemistry: Giving Students a Choice. *J. Chem. Educ.* **2012**, *89*, 1378–1383.
11. Kammler, D. C.; Truong, T. M.; VanNess, G.; McGowin, A. E. A Service-Learning Project in Chemistry: Environmental Monitoring of a Nature Preserve. *J. Chem. Educ.* **2012**, *89*, 1384–1389.
12. Hatcher, M. E.; Aragon, E. P. Combining Active Learning with Service Learning: A Student-Driven Demonstration Project. *J. Chem. Educ.* **2002**, *79*, 462–464.

13. Esson, J. M.; Stevens-Truss, R.; Thomas, A. Teaching Biochemistry at a Minority-Serving Institution: An Evaluation of the Role of Collaborative Learning as a Tool for Science Mastery. *J. Chem. Educ.* **2005**, *82*, 571–574.
14. Peters, A. W. Service-Learning in Introductory Chemistry: Supplementing Chemistry Curriculum in Elementary Schools. *J. Chem. Educ.* **2005**, *82*, 1168–1173.
15. LaRiviere, F. J.; Miller, L. M.; Millard, J. T. Showing the True Face of Chemistry in a Service-Learning Outreach Course. *J. Chem. Educ.* **2007**, *84*, 1636–1639.
16. Cartwright, A. Science Service Learning. *J. Chem. Educ.* **2010**, *87*, 1009–1010.
17. Saitta, E. K. H.; Bowdon, M. A.; Geiger, C. L. Incorporating Service-Learning, Technology, and Research Supportive Teaching Techniques into the University Chemistry Classroom. *J. Sci. Educ. Technol.* **2011**, *20*, 790–795.
18. Harrison, M. A.; Dunbar, D.; Lopatto, D. Using Pamphlets To Teach Biochemistry: A Service-Learning Project. *J. Chem. Educ.* **2013**, *90*, 210–214.
19. Theall, R. A. M.; Bond, M. R. Incorporating Professional Service as a Component of General Chemistry Laboratory by Demonstrating Chemistry to Elementary Students. *J. Chem. Educ.* **2013**, *90*, 332–337.
20. Glover, S. R.; Sewry, J. D.; Bromley, C. L.; Davies-Coleman, M. T.; Hlengwa, A. The Implementation of a Service-Learning Component in an Organic Chemistry Laboratory Course. *J. Chem. Educ.* **2013**, *90*, 578–583.
21. Burand, M. W.; Ogba, O. M. Letter Writing as a Service-Learning Project: An Alternative to the Traditional Laboratory Report. *J. Chem. Educ.* **2013**, *90*, 1701–1702.
22. Sewry, J. D.; Glover, S. R.; Harrison, T. G.; Shallcross, D. E.; Ngcoza, K. M. Offering Community Engagement Activities To Increase Chemistry Knowledge and Confidence for Teachers and Students. *J. Chem. Educ.* DOI: 10.1021/ed400495m
23. *Example Syllabi for Community-centered Chemistry Courses. Campus Compact*; http://www.compact.org/category/syllabi/chemistry/, accessed January 10, 2014.
24. *Program Models by Issue: Environmental Issues. Campus Compact*; http://www.compact.org/category/program-models/program-models-service-by-issue-environmental-issues/, accessed January 10, 2014.
25. *People for Puget Sound.* http://pugetsound.org, accessed January 10, 2014.
26. *Toxics Cleanup Program, Tacoma Smelter Plume*; Department of Ecology, State of Washington. http://www.ecy.wa.gov/programs/tcp/sites_brochure/tacoma_smelter/2011/ts-hp.htm, accessed March 1, 2014.
27. *Protecting Families from Arsenic and Lead in Soils. Public Health - Seattle & King County*; http://www.kingcounty.gov/healthservices/health/ehs/toxic/TacomaSmelterPlume.aspx, accessed March 1, 2014.
28. *EPA Method 3050B for the Acid Digestion of Sediment, Sludges, and Soils*; http://www.epa.gov/osw/hazard/testmethods/sw846/pdfs/3050b.pdf, accessed March 1, 2014.

29. *NPL Site Narrative for Lower Duwamish Waterway*; 2001. Environmental Protection Agency. http://www.epa.gov/superfund/sites/npl/nar1622.htm, accessed April 1, 2014.
30. Trim, H. *Restoring our River; Protecting our Investment: Duwamish River Pollution Source Control. Prepared for the Duwamish River Cleanup Coalition*; 2004. http://duwamishcleanup.org/wp-content/uploads/2012/02/DRCCSourceControlReport.pdf, accessed April 1, 2014.
31. Latch, D. E.; Whitlow, W. L.; Alaimo, P. J. Incorporating an environmental research project across three STEM courses: A collaboration between ecology, organic chemistry, and analytical chemistry students. In *Science Education and Civic Engagement: The Next Level*; Sheardy, R. D., Burns, W. D., Eds.; ACS Symposium Series 1121; American Chemical Society: Washington, D.C., 2012; pp 17−30.
32. *Poisoned Waters. PBS Frontline*; http://www.pbs.org/wgbh/pages/frontline/poisonedwaters/, accessed April 1, 2014.

Chapter 12

The LEEDAR Program: Learning Enhanced through Experimental Design and Analysis with Rutgers

David A. Laviska,* Kathleen D. Field, Sarah M. Sparks, and Alan S. Goldman

Department of Chemistry and Chemical Biology, Rutgers, The State University of New Jersey, 610 Taylor Road, Piscataway, New Jersey 08854
*E-mail: dlaviska@gmail.com.

Designing effective, enlightening experiments and following up with intelligent analysis of experimental results are critical components of productive research in the sciences. Unfortunately, most young scientists receive little introduction to the creative and analytical skills necessary for such endeavors prior to the start of graduate-level studies. While the science curricula of many secondary schools usually include well-researched and reliable classroom or laboratory experiments, these scripted exercises offer little insight into what it takes to be an independent researcher. The outreach program LEEDAR: Learning Enhanced through Experimental Design and Analysis with Rutgers was developed to expose high school students to the basic tenets of experimental design - a process through which they can independently form a hypothesis and then design, conduct, and present the results of their own unique classroom experiments. The central goal has been to forge connections between learning in the sciences at the high school, undergraduate, and graduate college levels, while creating an active-learning environment. In addition to encouraging young students to consider advanced study in the STEM disciplines, our program provides valuable service-learning experience to the graduate students who volunteer as mentors.

© 2014 American Chemical Society

Introduction

Traditionally, graduate programs in chemistry focus heavily on research and training students to function effectively as independent scholars (the scientific method, experimentation, analysis of results, publishing, etc.). Opportunities to learn about effective pedagogical practices and the communication of scientific concepts to less advanced students and/or the general public are often restricted to brief assignments as Teaching Assistants (TA's). Doctoral candidates who are not hired as TA's (due to alternate funding sources such as scholarships, grants, fellowships, etc.), typically gain no teaching experience at all prior to defending their research dissertations. While this lack of training may be of little detriment to newly minted Ph.D. chemists who will focus their careers in the research laboratory, it is a profound omission for those looking for jobs in academia and especially those who have a passion for teaching and communicating their enthusiasm for science.

Over the past seven years, graduate students and post-doctoral associates at Rutgers University (primarily, but not exclusively, affiliated with Primary Investigator Alan S. Goldman) have experimented with a series of service-based projects that provide in-depth teaching experiences, both in and out of the classroom. While the program is continually evolving, the basic format involves teams of graduate students meeting with science classes at local high schools with the dual goals of sharing information about advanced study and careers in the sciences and introducing the fundamental creativity involved in experimental design and scientific exploration. Now known as the LEEDAR Program (Learning Enhanced through Experimental Design and Analysis with Rutgers), this highly successful service-learning initiative provides a unique opportunity for graduate students to learn fundamental teaching skills, and perhaps more importantly, shows high school students that scientific research can be creative, fun, and an attractive, rewarding career option for them to consider as they think ahead toward life after high school.

The LEEDAR program was conceived and developed with funding from the National Science Foundation (NSF) under the auspices of the Center for Enabling New Technologies Through Catalysis (CENTC). CENTC was established in 2007 as an NSF Phase II Center for Chemical Innovation (CCI). With an emphasis on collaboration, CENTC brings together researchers (18 primary investigators and dozens of post-doctoral scholars and undergraduate and graduate students) from 14 North American universities and one national laboratory (Figure 1). The central scientific mission of the CCI is to collaboratively address economic, environmental and national security needs for more efficient, inexpensive and environmentally friendly methods of producing chemicals and fuels from a variety of feedstocks (*1*). Beyond fostering creative research in the chemical sciences, an equally important component of the CENTC mission includes training graduate students and postdoctoral scholars to communicate effectively with the general public about their fields of expertise. The advantages of this training are obvious: the students gain valuable insight into the skills required for effective science pedagogy, and the general public benefits through increased exposure to science and its practical applications. Since the goals of LEEDAR fit comfortably within the larger CENTC framework, our program has benefitted from continuous financial support from the NSF CCI program since its inception.

List of CENTC institutions by region

- West coast:
  - University of California, Berkeley
  - University of California, Santa Barbara
  - University of Washington

- East coast:
  - Massachusetts Institute of Technology
  - North Carolina State University
  - Rutgers, the State University of New Jersey
  - University of North Carolina, Chapel Hill
  - University of Ottawa
  - University of Rochester
  - Yale University

- North central:
  - University of Illinois, Urbana-Champaign
  - University of Michigan, Ann Arbor
  - University of Wisconsin, Madison

- South central:
  - Los Alamos National Laboratory
  - University of North Texas

Figure 1. Presentation slide showing CENTC institutions.

What follows is a condensed summary of the history, development, and propagation of the LEEDAR program. We believe the general model we have developed over the past seven years is a successful example of synergy between national organizations (NSF, CENTC), higher education (Rutgers University), and secondary education (various local high schools). While there have been many adjustments to the program over the years, the benefits of this type of service-learning outreach paradigm have been clear from the very beginning. We hope that chemistry departments at other colleges and universities will learn from our experiences and take the initiative to explore similar programs in the future.

History and Development of the LEEDAR Program

Over the past several decades, teachers at all levels of education, from elementary schools, upward through middle and high schools and universities, have expended considerable effort and creativity toward the goal of stimulating their students' interest in studying the physical sciences. While young children often display unabashed creativity coupled with curiosity and fearlessness, older students seem less inclined to explore purely for the sake of discovery. In an effort to reignite their sense of wonder, dozens of flashy demonstrations of chemical phenomena have been developed for use both in and out of the classroom (*2–8*). Fortunately, chemistry is a tremendously diverse field of study; exciting experiments of all sorts are simple (and safe) to execute, and the results (e.g., color changes, mini explosions, noises, generation of gases, etc.) are sufficiently entertaining to generate enthusiasm among audience members, if not necessarily genuine interest in further study of the science *behind* the experiments. In general, observers gain little or no understanding of the chemical principles that underlie the experiments, and therefore, how the methodologies were developed and why they were successful in the demonstration setting. In this context, the fundamental precepts of the scientific method are neglected entirely; a failed experiment can be enormously instructive in the research laboratory, but will almost certainly undermine the entertainment value of a "demo show".

Authors of an impressive breadth of articles have thoroughly communicated the value of "active" vs. "passive" learning through detailed studies in a variety of instructional contexts (9–27). We were fortunate to benefit from their knowledge, and decided to avoid any approach that encourages passive learning as we designed our preliminary outreach initiative. In its most traditional sense, passive learning involves a "scientific expert" who lectures to students in order to disseminate information that needs to be memorized and then regurgitated by students during assessment (i.e., exams). Typically, the instructor varies minimally from a pre-set curriculum, and there is little opportunity for students to display their curiosity through questions and interaction with the instructor or their colleagues. Students tend to take notes verbatim, memorize material regardless of comprehension, and avoid self-evaluation. Unfortunately, this leads to a process in which students endeavor to learn enough to meet a grade standard rather than seeking an in-depth understanding of chemical *concepts*. On the contrary, in an active learning environment, the instructor facilitates and guides the learning process rather than handing out answers, and acts as an interactive mentor rather than an unapproachable "expert". Flexible curricula and lesson plans allow for and encourage student participation and interaction, including asking questions and helping or teaching other students. Ideally, students accrue knowledge through a process of inquiry and experimentation, and ultimately, take significantly greater personal responsibility for their individual learning experiences.

Preliminary Goals

As we began to formulate the preliminary plans for a new "outreach" program at Rutgers University, our primary goal was simply to create connections between high school students and higher education. That is, we sought to shed light on the reasons why people choose to study and prepare for careers in the sciences. Despite our society's well-publicized addiction to information and electronics, high school students remain remarkably ignorant of the demands they will face as undergraduate students and beyond, should they choose these advanced educational options. Unfortunately, young students still cling to outdated clichés about scientists: they are likely to be intelligent, but also socially insular, awkward, unattractive, and boring (i.e., "geeks"). Furthermore, young people seem to have strongly-held, preconceived ideas about the study of science itself: it is boring, difficult, and tedious; it involves extensive memorization; it's only for "smart" people, etc.

Recognizing these biases, we decided to formulate a program based around interactive discussions of "science in the news" – i.e., science-related topics with which we felt certain the students had some familiarity, based on their exposure to publicity through popular media outlets. Eventually, we settled on the themes of Global Warming and Climate Change. Most high school students in this country are highly aware of "global warming", since it has permeated popular media with daily stories about rising global temperatures, greenhouse gases, energy conservation, and "green" technology. However, despite frequent media coverage, students (and indeed, laypeople in general) have little understanding of the underlying causes of the proposed environmental changes. As a result, the topics of global warming, carbon dioxide emissions, and natural resources lead to lively discussions that high school students can be interested in while learning the science behind the news articles. For example, discussions of the greenhouse effect often elicit comments from students about damage to the atmospheric ozone layer – an almost entirely unrelated phenomenon (*28*). Using relatively few visual aids (slides, diagrams), our team focused on wide-ranging discussions with the students and attempted to make personal connections with as many of them as possible – all in the context of their usual classroom, and in cooperation with their science teachers. Although the exact content of the discussions varied between classes, we attempted to adhere to one or more of the following curricular guidelines:

- What causes global warming?
- What are fossil fuels?
- What are greenhouse gases?
- Why does CO_2 act as a greenhouse gas?
- Where does CO_2 come from?
- What are the top energy-related problems in the world today?
- Why are we so dependent on fossil fuels?
- What are the environmental ramifications of climate change?

Dry Ice as a Learning Tool

As part of our first outreach sessions, we took dry ice (solid CO_2) into the classrooms and talked about the basic structure and chemistry of the carbon dioxide molecule in very simple terms (*29, 30*). The connection to the theme of global warming was obvious, but what we did not expect was the excitement something as simple as dry ice could generate among high school chemistry students. In fact, while dry ice is commonplace for most university researchers, it is rarely found in high school classrooms except under very tightly controlled circumstances, due to the obvious safety concerns (i.e., low sublimation temperature). Therefore, the novelty and unique physical properties of solid CO_2 guaranteed the attention of most students – particularly when we allowed them to handle it themselves (with appropriate personal protective gear: gloves and goggles). Since CO_2 is a relatively simple molecule, our team was able to discuss chemical concepts such as atoms, bonding, dipole moments, ideal gases, etc., depending on the level of student expertise and what concepts they had already learned in the context of their class curriculum. Whenever possible, we discussed one particularly striking example of human-generated CO_2 emissions: combustion from automobile engines (Figure 2).

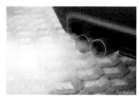

> **How much CO_2 is produced by burning gasoline?**
>
> • What is combustion?
>
> • What is the chemical reaction if gasoline = octane?
>
> • Let's balance the equation:
> $$2\ C_8H_{18} + 25\ O_2 \longrightarrow 16\ CO_2 + 18\ H_2O$$
>
> • Concepts:
>   • Stoichiometry, Molar Mass, Dimensional Analysis, Unit conversion, Density, Phase changes, Significant Figures

Figure 2. Presentation slide for discussion of automobile CO_2 emissions.

Fossil Fuels

Beyond the most obvious use for fossil fuels – i.e., as actual fuels for combustion in transportation and heating – students were generally unaware of the multitude of consumer products that are derived in whole or in part from non-renewable, fossil fuel sources. Since we are located in the state of New Jersey, where refineries and chemical manufacturing facilities are plentiful, discussions of fossil fuel-based consumer products were both interesting to the students and pertinent to their communities. These discussions served several important purposes. First, the students learned about one aspect of our nation's dependency on fossil fuels by discovering the tremendously broad array of items they use *daily* that are made from fundamental chemical building blocks derived from crude oil. Second, they were able to gain a better sense of scale concerning how many scientists of all kinds are employed by the combined chemical and fuel industries. Third, they were able to contextualize conservation efforts such as recycling plastics and gain a better understanding of both the chemistry of polymers and why recycling is feasible (in terms of chemical synthesis) and valuable (recapturing non-biodegradable waste and reusing the fundamental chemical building blocks). Figure 3 shows a few of the many slides used as a launching point for discussions of fossil fuel-derived consumer products. Those included here refer specifically to plastics and their uses and recycling.

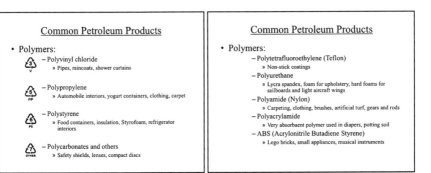

Figure 3. Presentation slides for discussion of consumer products derived from fossil fuels.

Experimentation

Reaching beyond simple discussions and hands-on experience with dry ice, we thought that introducing lab experiments into our interactive sessions would be an obvious way to engage the students while communicating fundamental chemical principles. Reports from many authors agree that active experimentation is one important tool for encouraging young people to seek out higher levels of understanding in the sciences (*31–34*). At first, we implemented a simple experiment found in many high school classrooms: the determination of the molar volume of an "ideal gas" (*35*, *36*). One of the most attractive aspects of this "laboratory experiment" is that no laboratory is required! Thus, students without access to bench space or specialized instructional classrooms could still carry out the experiment and achieve moderately accurate results. In addition, the experimental materials are cheap, reliable, and reasonably safe. Using balloons and dry ice, students were able to determine the molar volume of carbon dioxide by subliming a known mass of dry ice inside a balloon, taking measurements of the inflated balloon, calculating its volume, and performing the necessary calculations to determine the molar volume (Figure 4).

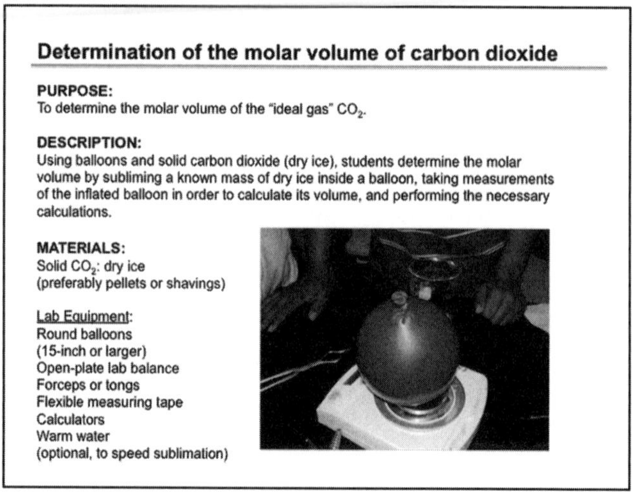

> **Determination of the molar volume of carbon dioxide**
>
> **PURPOSE:**
> To determine the molar volume of the "ideal gas" CO_2.
>
> **DESCRIPTION:**
> Using balloons and solid carbon dioxide (dry ice), students determine the molar volume by subliming a known mass of dry ice inside a balloon, taking measurements of the inflated balloon in order to calculate its volume, and performing the necessary calculations.
>
> **MATERIALS:**
> Solid CO_2: dry ice
> (preferably pellets or shavings)
>
> <u>Lab Equipment</u>:
> Round balloons
> (15-inch or larger)
> Open-plate lab balance
> Forceps or tongs
> Flexible measuring tape
> Calculators
> Warm water
> (optional, to speed sublimation)

Figure 4. Preliminary laboratory experiment: Molar volume of carbon dioxide.

Although this experiment is well known and used in many general chemistry classrooms and even for demonstration purposes, the students' lack of connection between the ideas addressed in our preliminary discussions and the chemical concepts pertinent to the experimental procedure quickly became apparent. Typically, only one student per group did the majority of the work in collecting data, meaning the others were no longer *actively* participating, but had reverted to *passive* observation. In addition, since the desired answer is "known" (~22.4 liters/mol at standard temperature and pressure), students had little incentive to make accurate measurements. But perhaps most damaging to the goals of our program, little or no creativity was required of the students: a detailed procedure was given to the students and therefore, they didn't need to think about how to solve a problem. In short, the "experiment" wasn't *experimental* at all; it was merely an exercise in following instructions. While observing the students conducting this "cookbook" experiment, we drew several valuable conclusions about how to change our program in order to more effectively promote creativity and active learning.

Experimental Design

Revisiting our goals for this outreach initiative, it became obvious that we needed to avoid traditional laboratory experiments as designed for the high school classroom (*37*). While there is no question that simple, well-designed, safe, pedagogically sound experiments are a critical component of introductory science classes, our goal was to show high school students the beauty and excitement of science as it exists *outside of textbooks*. It is easy to have the opinion that science is exciting and creative once a sufficient level of expertise has been attained that allows for individual, novel experimentation and discovery. In general terms, this is what researchers do: they build on existing concepts and data, but design *new* experiments, in order to collect *new* data that leads to a *new* understanding of the concepts being studied.

Therefore, after implementing experiments with detailed procedures over the first two years of our program, we changed our strategy and began to implement a new interactive module called "Design your own experiment" (*38*). This new module allowed the student to become the "researcher," facilitating an active learning environment full of creative exploration from formation of the preliminary hypothesis, through the experimental design, to the final data analysis and reporting of results. In essence, our goal became to demonstrate the most fundamental aspect of the scientific method, i.e., designing an experiment to test a hypothesis. We wanted to develop young students' ability to solve problems involving simple chemical concepts that are inter-related in a relatively complex "real world" fashion, while also encouraging them to seek explanations of physical observations from within their own knowledge base and intuition. Recognizing that posing a single question such as "What is the molar volume of an ideal gas?" would be too restricting and very likely lead to repetitive results, we vastly broadened the scope of the experimental module and posed several very general questions (Figure 5).

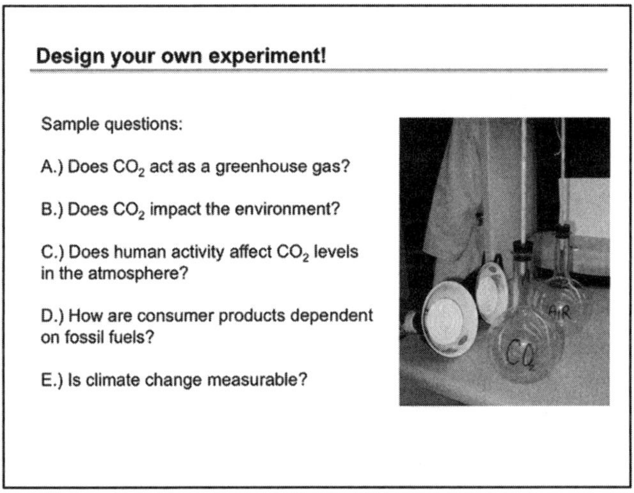

Design your own experiment!

Sample questions:

A.) Does CO_2 act as a greenhouse gas?

B.) Does CO_2 impact the environment?

C.) Does human activity affect CO_2 levels in the atmosphere?

D.) How are consumer products dependent on fossil fuels?

E.) Is climate change measurable?

Figure 5. General questions relating to the experimental design module.

These questions were purposefully vague; the intent was to encourage the students to think independently about formulating a hypothesis that they might reasonably be able to confirm or refute through simple experiments. While our team had concerns about the potential for chaotic results, the new "Design your own experiment" module was immediately successful (*vide infra*) and our interactive program was given a formal name: Learning Enhanced through Experimental Design and Analysis with Rutgers: the LEEDAR Program (*39*).

LEEDAR Program - 2014

Overview

Beginning in September 2014, the Rutgers University chemistry service-learning outreach program, now called the LEEDAR Program, began its eighth consecutive academic year of service to the physical science classes in secondary schools of nearby communities (*40, 41*). Under the auspices of this program, teams of graduate students, postdoctoral associates, and undergraduates from the chemistry department at Rutgers (all unpaid volunteers) go to local high schools in order to meet with students in science classes and discuss relevant, real-life applications of scientific concepts. LEEDAR team members learn extensively about teaching and communicating science through service to local high school teachers and their classes, while acting as mentors to the younger students. The LEEDAR curriculum can be tailored to the individual needs of the high school teachers, with one-day, two-day, three-day, or even longer commitments available, depending on the scope of the program as mutually agreed upon with the high school teacher and administration.

The budget for our program is modest, since very few supplies are needed for the classroom visits (e.g., dry ice, safety supplies, balloons, etc.). As such, the primary resource needed for running the LEEDAR program is *time*, not money. Typically, the program involves several days of interaction, spread over several weeks (or months), with a strong emphasis on the "Design your own experiment" teaching module. However, shorter presentations (even as brief as a single visit) can be arranged if a particular teacher is new to our program or has limited flexibility for scheduling within the standard school calendar format. Regardless of the total number of LEEDAR visits, presentations are designed and scheduled to fit within the preexisting class time grid as established by the high school administration. Whenever possible, visits are planned to coincide with lab periods, since these are typically longer than non-lab sessions, in order to maximize contact time for the students. Volunteers on the LEEDAR team participate according to their individual schedules and availability. During an average year, the scheduling process is completed for LEEDAR sessions with participating high schools in early fall, if possible. This allows LEEDAR volunteers to plan ahead for time away from classes and/or research over the remainder of the academic year. A general description of the program along with an explanation of the logistical concerns follows below.

Phase One: Introduction and Discussion

Prior to the first visit to a high school, a LEEDAR coordinator settles dates and the scope of the program with an interested high school teacher. The first day of interaction between LEEDAR volunteers and high school students adheres closely to the paradigm described earlier in this article. There is an interactive discussion focusing on the topics of energy, fossil fuels, and renewable fuel sources based on a question-answer (inquiry-based) format rather than a simple lecture. These wide-ranging discussions typically culminate in a discussion of the role of carbon dioxide as a greenhouse gas and the introduction of the students to CO_2 in its solid form: dry ice. The novelty of working with dry ice is one way the Rutgers team brings the "university experience" into the high school: the students are excited to have the opportunity to observe and handle a material that is not usually available in the high school laboratory. A short list of in-class activities enable the LEEDAR personnel to review basic chemical concepts (stoichiometry, gas laws, phase changes) while helping the students make connections between these fundamental topics and phenomena they have heard of outside the classroom, such as climate change, the greenhouse effect, etc. As mentioned briefly above, one of the in-class activities is balancing the equation for combustion of octane and then asking students to calculate the amount of CO_2 produced during this reaction, starting from one gallon (~4 liters) of gasoline or octane given the density and molecular weights. Natural and man-made sources of CO_2 emissions are discussed, as is the manner in which CO_2 interacts with infrared radiation and why this might lead to a warming trend in our atmosphere. Another portion of the discussion encompasses the origins of fossil fuels, how they are processed, their uses as chemical building blocks, and why having a limited supply of natural resources is a potential cause for concern. Significantly, LEEDAR personnel are not tasked with arguing *for* or *against* the existence of global warming, but rather the goal is to inform the students and encourage them to question the *scientific evidence* while forming their own conclusions. At the conclusion of the discussion, a series of questions are presented to the students and they are invited to brainstorm questions that may lead to testable hypotheses for which they could potentially design and execute their own unique experiments.

Phase Two: Experimental Design – Ideas, Materials, and Equipment

After the preliminary visit from LEEDAR personnel, the students begin the design and development phase for their individual experiments. Usually working in pairs, the students adhere to incremental deadlines as set forth by their teacher, in conjunction with the LEEDAR team. Examples of critical components of the design process are as follows:

- Title (basic idea for the experiment);
- Hypothesis (should be well-defined and testable);
- Materials (must be available and safe for the classroom);
- Procedure (should be of appropriate scale);
- Plans for reporting results (poster vs. oral);
- Date(s) for experimentation and number of trials planned; and,
- Dates for summary and presentation of results.

Importantly, most of the creative control is left in the hands of the students, although they are required to seek some practical guidance from both the teacher and LEEDAR personnel. While creativity is encouraged, concerns such as cost, safety, and physical scale of the experiments are closely monitored as the students submit their ideas and plans according to the timeline discussed above. At each stage, assessment by the teacher can (and often does) call for revisions from the students. Ultimately, the students take responsibility for the learning process, meet deadlines, and engage with the concepts at a personal level unique to each individual.

Phase Three: Experimentation

Once the planning phase for the project has been completed, the students test their hypotheses through experimentation, collect and analyze results, and draw their own conclusions. As stated previously, the curricula can be customized, with one-, two-, or three-day programs available, and so subsequent visits with LEEDAR personnel can be planned as desired by the students and accommodated within the teacher's lesson plans. The additional visits usually entail a planning day in which students have the opportunity to ask their LEEDAR mentors questions about their setup, materials, procedure, and concerns before they conduct their experiment. If desired, LEEDAR members can also witness the experiments in action and help the students analyze their data and revise their experiments for additional trials. In general, the students devote a significant amount of time and effort to the process of experimental design, and are therefore excited about explaining and showing off their results, even with mistakes and unexpected pitfalls. Their discovery-based enthusiasm is easily observed in the care and attention to detail they invest in the selection and arrangement of rudimentary lab equipment. Several examples are shown in Figure 6.

At the outset of the design and development process, many students are intimidated by the lack of a prescribed methodology. However, by the time they are conducting experiments to test their hypotheses, most have developed a strong appreciation for the freedom and personal creativity involved in setting up an experiment to answer a question they themselves have asked. Assessment of student opinions has yielded a remarkable body of comments that illuminate the shift in attitudes throughout their progression from preliminary brainstorming to final conclusions. The following quote from a student illustrates this point: *"The lack of handholding was interesting, considering nearly every lab we do in class is very guided and controlled. The idea of answering questions of our own or creating things we've never had the opportunity to do was exciting."* Students also seem to appreciate the ability to make connections with issues from outside the classroom: *"It was interesting to be able to have freedom to come up with your own experiment and connect it to the real world."*

Phase Four: Presentation of Results and Evaluations

At the conclusion of the experimental portion of the program, the students compile and present their results (including a thorough description of the experimental details) to the class, teachers, and LEEDAR team. These presentations have taken many forms, but the most successful presentation days have been comprised of poster and/or oral presentations. Some teachers have allowed students to present in video or web formats that do not require a live audience. However, we have found that this type of presentation effectively undercuts the mentor/student relationships that were developed throughout the multi-day program. Poster and oral presentations truly mimic the process through which "real world" researchers communicate their results to the scientific community (e.g., ACS conferences). In the process of presenting experimental results to their peers and mentors, the students gain valuable experience with essential communication skills, and they become the "expert" - taking control of what their fellow classmates learn from their project. They also take responsibility for their work, actively learning about a topic rather than memorizing ideas without full comprehension. According to one student: *"I realized that creating experiments is a great way to learn about a topic."*

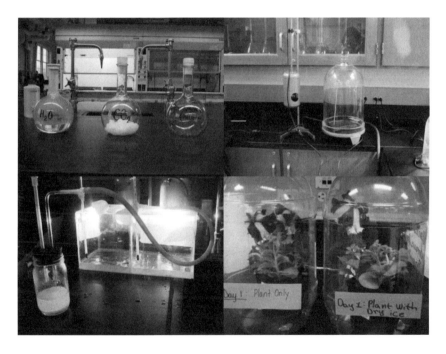

Figure 6. Examples of student experiments.

In addition to reviewing their results, we try to allow time for question and answer sessions so students can ask about what to expect in college, courses to take, careers, future opportunities in science, etc. As one student remarked: *"It showed how there are jobs out there that work with experiments on a daily basis. Therefore, I hope to study chemistry in the future and obtain a job like that."* The teachers also find the LEEDAR mentor/student relationship beneficial in helping their students prepare for the world beyond high school: *"There is a 'freshness' having someone from the outside come in and share with my students; they serve as role models of what will be expected of them in college."*

As mentioned in the introduction to this paper, LEEDAR team members also hope to show that science is important to society and that it can be a fun and exciting course of study. By this stage of our program, we hope that the clichéd image of the geeky scientist in a lab coat and goggles has been largely dispelled. Through the process of discovery facilitated by the LEEDAR program, many students begin to realize that being an "expert" is not a requirement in order to have fun with learning and experimenting. As one student wrote on her evaluation: *"Someone doesn't have to be a scientist to create and conduct an experiment."*

Summary

Service-Learning through Mentoring

From our first impulse to reach out to the community around Rutgers University, through the founding and development of our outreach initiative, to the current iteration of the highly successful LEEDAR program as described in this article, the potential benefits to high school students and their teachers have been paramount on our list of goals. However, as the program has aged and evolved, another profound benefit has emerged that is critical to the propagation of our program and its continued growth. The undergraduates, graduate students and post-doctoral associates that perform service to the community by volunteering as LEEDAR mentors receive valuable training in communication and pedagogical skills. As mentioned in the introduction to this article, pedagogical training for university-level science students is extremely limited at most institutions. The LEEDAR program provides an ideal service-learning opportunity for these students, giving them experience with a broad spectrum of skills in a flexible, multidisciplinary context. A partial list of these learning experiences includes

- Lecturing and discussion-based teaching;
- Experimental design;
- Extemporaneous explanations of all sorts of scientific phenomena;
- Classroom dynamics;
- Collaborative teaching with high school teachers;
- Mentoring;
- Assessment with thoughtful, constructive feedback;
- Planning, scheduling, and curricular flexibility; and,
- Writing and editing.

With the exception of the first two items in the list above, graduate students typically receive little, if any, training in these areas. For those intending to make their careers in academia, the remaining skill sets can be profoundly beneficial.

Each year, the students who have participated as LEEDAR team members confirm that the experience has value far beyond what they originally expected: *"I got involved* [with the LEEDAR program] *because I like to work with kids. In the final analysis, though, I think I learned as much or more than the high school students. This has been BY FAR the most rewarding teaching experience I have had throughout my education. Since I hope to become a professor of chemistry, my affiliation with LEEDAR has been invaluable."*

As the importance of the service-learning component of the LEEDAR program has crystallized in recent years, evaluation by and feedback from the volunteer mentors has become an increasingly high priority. While the benefits are frequently discussed among team members, to date, no formal assessment process has been instituted. As of 2014, a proposal to the chemistry department at Rutgers University is in development that suggests enumeration of volunteer hours that could lead to potential value for the graduate students including course credit and/or formal recognition through letters of acknowledgment or departmental awards.

LEEDAR Program Challenges and Pitfalls

Over the past six years, we have experienced a significant number of challenges that have required changes to the program, resulting in a fluid, ever-evolving format. No two schools are identical – and the same is true for individual high school teachers and of course, students.

Students often find time constraints to be a major drawback to the program. Teachers have limited flexibility in their state-mandated curricula, and therefore, scheduling a multi-day program such as LEEDAR can be a major challenge. With strict time constraints, students must adhere to incremental deadlines, as discussed above, and failure to do so can result in poorly planned experiments. Students sometimes do not have the time to repeat or revise their experiments, and one runthrough of the procedure may not be sufficient. Deadlines vary between different high schools and teachers, requiring a highly flexible approach to scheduling by the LEEDAR team. Teachers must clearly and carefully define the guidelines for presentations/reports and students are restricted to materials that can be easily found in the classroom/laboratory or purchased by their parents. With both time and cost constraints, students must be highly organized, use their time wisely, and choose materials and equipment with care. It comes as no surprise that this is a major challenge for young, inexperienced students, and therefore, the teacher and LEEDAR mentors must play active, leading roles. It is very common that proposed experiments are too ambitious, and without proper guidance, the students will fail on a logistical level, and ultimately miss the central message about fun and creativity in science.

Among the cognitive challenges faced by the students during the learning process is how to deal with the frustration of a "failed" experiment. Additionally, students can work hard, but still not understand how to support a hypothesis. Making the connection between the real world and/or the relevant curricular concepts to the chosen experiments can also be difficult. It is important for the

LEEDAR team members to establish the idea that every experiment is a learning experience, whether or not an answer is found for the initial question, and that science involves constant exploration and revision of hypotheses. These types of pitfalls not only teach high school students the difficulties behind planning and executing experiments, but they are opportunities for the teacher and LEEDAR team members to grow as educators.

Finally, the skills needed for effective presentation of the students' experimental results are often lacking due to their relative youth and inexperience. Here again, the teacher and the LEEDAR mentors must play a strong role in guiding the process and teaching the students the value of simple, clear communication and understanding how to explain a topic to an audience unfamiliar with their project.

Feedback

At the conclusion of the LEEDAR program, we ask the students and teachers to fill out voluntary, anonymous evaluation forms in an attempt to assess the value of the program. At the beginning of every new academic year, the LEEDAR team reviews evaluations from the prior year in order to make changes when possible and to stay informed of potential pitfalls and areas that may need improvement. While we do have a numeric rating system, the most illuminating feedback generally comes from the individual hand-written comments. Several student and teacher comments have already been included in this article (*vide supra*). In general, we have been gratified to see that students often seem to grasp the concepts we attempt to communicate through the LEEDAR program:

- *"I learned that even when your hypothesis is incorrect you still didn't fail because you can always learn something from the experiment."*
- *"Although we did not get the results that we hoped for, my group and I were still happy as it gave us the experience that an actual scientist goes through."*
- *"I learned that a lab without a procedure in a textbook is much harder than it seems."*
- *"This was a very fun project, much more so than the labs where we simply follow instructions. I enjoyed coming up with the procedure and working with my group a lot."*
- *"There were times when I felt like giving up but I convinced myself to keep going. But this experience made me more prepared for the real challenges a scientist might face."*

One unexpected outcome from the LEEDAR Program involved the pedagogical insight gained by the *teachers* as they guided their students through the process and then observed the final presentations. As the following quote illustrates, teachers have been able to learn from our program and even modify their own curricula in order to more effectively guide and assess their students:

- *"Participating with LEEDAR taught me that my students did not feel comfortable speaking about their experiments; when a student has to make a presentation of lab results, it becomes very clear if they don't understand the concepts. As a result, I changed my entire approach to lab assignments for my year-long curriculum, and selected one group per week to present their results to the entire class. This experience gave the students valuable practice in presenting technical material and answering questions posed to them. It also gave the class much needed 'closure' to their lab assignments. As a teacher, the LEEDAR experience provided invaluable pedagogical insight. I look forward to making your program an annual event in my classroom."*

Growth of the Program

The preliminary program was launched in 2007, and had reached over 2,000 students by the end of the 2012-13 academic year. The breakdown of the number of schools, teachers, classes, and students that have been involved is provided in the table shown in Figure 7. The number of teachers and schools involved with the LEEDAR program fluctuates each year depending on the teachers' class assignments, curricular schedules, approval from the individual school administrative personnel, and availability of LEEDAR volunteers. Our initial goal was to grow the program each year, in order to serve as many local high school students as possible. However, it is clear from the data, that there is a maximum level of community involvement – limited by LEEDAR resources and time constraints. All LEEDAR volunteers are full-time students or research associates, and therefore scheduling is challenging for all parties concerned in this outreach effort. At the present time, we are satisfied with visiting 15-20 classes per academic year.

	2007-2008	2008-2009	2009-2010	2010-2011	2011-2012	2012-2013	Totals
Schools	1	3	4	2	4	2	5[a]
Classes	3	16	17	15	14	21	86
Teachers	1	5	6	4	7	6	8[b]
Students	75	345	400	450	400	450	2120

Figure 7. Statistics for the LEEDAR Program from 2007-2013. (Notes: Data from the 2013-2014 academic cycle are not yet available. [a]Since many schools participate in the LEEDAR program more than once, this figure represents the total number of unique schools visited. [b]This figure represents the total number of unique high school teachers who have participated in the LEEDAR program.)

Future Directions

The LEEDAR program continues to stay focused on expanding by connecting with more high schools and teachers to participate as well as gaining new volunteers to mentor students. Several alternative curricula are in development so that we can diversify the subjects we discuss in the preliminary visits as well as to encourage a greater breadth of student-designed experiments. Some ideas we have considered include alternative fuels, nanotechnology, green chemistry, and environmental custodianship. We have also initiated an application-based summer research program in which high school students learn and work in research laboratories at Rutgers University. In addition to giving them a true hands-on laboratory experience, the summer program allows students to establish deeper personal and professional connections with both graduate students and professors. The central goal of our program remains the same: forge new connections and communicate the fun and creativity of scientific exploration.

Conclusions

The LEEDAR program is an interactive community outreach program that connects high school students with professional and academic scientists, while providing a valuable service-learning opportunity for graduate students in the sciences. Our unique approach includes a "Design your own experiment" module that demonstrates the fundamental aspects of the scientific method and develops young students' ability to solve problems involving simple chemical concepts that are inter-related in a relatively complex "real world" fashion. The LEEDAR program fosters an active learning environment: students to take control of their learning, and are responsible for at least part of every phase of the process from preliminary ideas, through planning and execution, to the presentation of final results. LEEDAR team members act as mentors while encouraging high school students to see the exciting potential of careers based in the sciences. Finally, by serving the community as LEEDAR volunteers, graduate students, research associates, and undergraduates gain invaluable training as teachers and in the communication of chemistry and the sciences in general.

Acknowledgments

This work was supported by the National Science Foundation under the CCI Center for Enabling New Technologies Through Catalysis, CHE-0650456 and CHE-1205189. The authors would like to thank the students of Highland Park (Highland Park, NJ), Bridgewater-Raritan (Bridgewater, NJ), Ridge (Basking Ridge, NJ), Colonia (Colonia, NJ), and Westfield (Westfield, NJ) high schools and the teachers that collaborated with the authors over the past seven years, including James Danch, Mabel Hyunh, Benjamin Lee, Judith McCloughlin, Rosemary Mead, Srividhya Periyaswamy, Keisha Stephen, and Carol Wenk.

References

1. University of Washington. *Center for Enabling New Technologies through Catalysis Home Page*; http://depts.washington.edu/centc (accessed September 20, 2014).
2. Bodsgard, B. R.; Johnson, T. A.; Kugel, R. W.; Lien, N. R.; Mueller, J. A.; Martin, D. J. *J. Chem. Educ.* **2011**, *88*, 1347–1350.
3. Gammon, S. D. *J. Chem. Educ.* **1994**, *71*, 1077–1079.
4. Kerby, H. W.; Cantor, J.; Weiland, M.; Babiarz, C.; Kerby, A. W. *J. Chem. Educ.* **2010**, *87*, 1024–1030.
5. Ophardt, C. E.; Applebee, M. S.; Losey, E. N. *J. Chem. Educ.* **2005**, *82*, 1174–1177.
6. Swim, J. *J. Chem. Educ.* **1999**, *76*, 628–629.
7. Voegel, P. D.; Quashnock, K. A.; Heil, K. M. *J. Chem. Educ.* **2004**, *81*, 681–684.
8. Waldman, A. S.; Schechinger, L.; Norwock, J. S. *J. Chem. Educ.* **1996**, *73*, 762–764.
9. Brooks, B. J.; Koretsky, M. D. *J. Chem. Educ.* **2011**, *88*, 1477–1484.
10. Burke, K. A.; Greenbowe, T. J.; Lewis, E. L.; Peace, G. E. *J. Chem. Educ.* **2002**, *79*, 699.
11. Cotes, S.; Cotua, J. *J. Chem. Educ.* **2014**, *91*, 673–677.
12. Hageman, J. H. *J. Chem. Educ.* **2010**, *87*, 291–293.
13. Hatcher-Skeers, M.; Aragon, E. *J. Chem. Educ.* **2002**, *79*, 462–464.
14. Hinde, R. J.; Kovac, J. *J. Chem. Educ.* **2001**, *78*, 93–99.
15. Hodges, L. C. *J. Chem. Educ.* **1999**, *76*, 376–377.
16. Katz, M. *J. Chem. Educ.* **1996**, *73*, 440–445.
17. Kovac, J. *J. Chem. Educ.* **1999**, *76*, 120–124.
18. LaRiviere, F. J.; Miller, L. M.; Millard, J. T. *J. Chem. Educ.* **2007**, *84*, 1636–1639.
19. Nee, M. J. *J. Chem. Educ.* **2013**, *90*, 1581–1582.
20. Oliver-Hoyo, M. T.; Allen, D.; Hunt, W. F.; Hutson, J.; Pitts, A. *J. Chem. Educ.* **2004**, *81*, 441–448.
21. Paulson, D. R. *J. Chem. Educ.* **1999**, *76*, 1136–1140.
22. Penn, R. L.; Flynn, L.; Johnson, P. *J. Chem. Educ.* **2007**, *84*, 955–960.
23. Phipps, L. R. *J. Chem. Educ.* **2013**, *90*, 568–573.
24. Ross, M. R.; Fulton, R. B. *J. Chem. Educ.* **1994**, *71*, 141–143.
25. Shaver, M. P. *J. Chem. Educ.* **2010**, *87*, 1320–1323.
26. Steiner, R. P. *J. Chem. Educ.* **1980**, *57*, 433–434.
27. Worrell, J. H. *J. Chem. Educ.* **1992**, *69*, 913–914.
28. Kerr, S. C.; Walz, K. A. *J. Chem. Educ.* **2007**, *84*, 1693–1696.
29. Laviska, D. A.; Stephen, K.; Sparks, S. M.; Pelczar, E. M.; Spink, R.; Goldman, A. S. Building a bridge between high school education and CENTC research: An interactive program for integrated outreach, mentoring, and education on greenhouse gases and global warming. Abstracts of Papers, 237th ACS National Meeting, Salt Lake City, UT, United States, March 22−26, 2009; CHED-1206.

30. Laviska, D. A.; Stephen, K.; Sparks, S. M.; Pelczar, E. M.; Spink, R.; Goldman, A. S. Designing outreach curricula to focus on mentoring high school students: The relationship between greenhouse gases and global warming. Abstracts of Papers, 237th ACS National Meeting, Salt Lake City, UT, United States, March 22−26, 2009; CHED-067.
31. Chan, Y. M.; Hom, W.; Montclare, J. K. *J. Chem. Educ.* **2011**, *88*, 751–754.
32. Heinze, K. F.; Allen, J. L.; Jacobsen, E. N. *J. Chem. Educ.* **1995**, *72*, 167–169.
33. Long, G. L.; Bailey, C. A.; Bunn, B. B.; Slebodnick, C.; Johnson, M. R.; Derozier, S.; Dana, S. M.; Grady, J. R. *J. Chem. Educ.* **2012**, *89*, 1249–1258.
34. Thomas, C. L. *J. Chem. Educ.* **2012**, *89*, 1259–1263.
35. Jackson, B. A.; Crouse, D. J. *J. Chem. Educ.* **1998**, *75*, 997–998.
36. Zaborowski, L. M. *J. Chem. Educ.* **1972**, *49*, 361.
37. Merritt, M. V.; Schneider, M. J.; Darlington, J. A. *J. Chem. Educ.* **1993**, *70*, 660–662.
38. Laviska, D. A.; Sparks, S. M.; Stephen, K.; Goldman, A. S. Evolution of an innovative university-to-high school outreach program: The importance of mentoring in encouraging students to think about careers in chemistry. Abstracts of Papers, 238th ACS National Meeting, Washington, DC, United States, August 16−20, 2009; CHED-026.
39. Laviska, D. A.; Sparks, S. M.; Goldman, A. S. Introducing the LEEDAR program - Learning Enhanced through Experimental Design and Analysis with Rutgers: Giving high school students some insight into what it means to be a scientist. Abstracts of Papers, 240th ACS National Meeting, Boston, MA, United States, August 22−26, 2010; CHED-94.
40. Laviska, D. A.; Field, K. D.; Goldman, A. S. Shedding light on a well-worn teaching paradigm: Asking students to think about how to design a successful experiment and watching them learn why research in chemistry is valuable, creative, and fun. Abstracts of Papers, 243rd ACS National Meeting, San Diego, CA, United States, March 25−29, 2012; CHED-1641.
41. Laviska, D. A.; Field, K. D.; Goldman, A. S. High School outreach staged demonstrations: Benefits and pitfalls of introducing young students to the process of experimental design. Abstracts of Papers, 244th ACS National Meeting & Exposition, Philadelphia, PA, United States, August 19−23, 2012; CHED-25.

Editors' Biographies

Elizabeth S. Roberts-Kirchhoff

Elizabeth Roberts-Kirchhoff is Associate Professor of Chemistry and Biochemistry at the University of Detroit Mercy. Her research interests include the mechanism of action of cytochrome P450 enzymes; the analysis of metals in food and health supplements including kelp, clay, and protein powders; and the analysis of pesticides in water.

Roberts-Kirchhoff received a B.S. in Chemistry from Texas A&M University and a Ph.D. in Biological Chemistry from the University of Michigan. After postdoctoral research at Wayne State University and The University of Michigan, she joined the faculty at the University of Detroit Mercy in 1997.

Matthew J. Mio

Matthew Mio is an Associate Professor at the University of Detroit Mercy in the Department of Chemistry and Biochemistry. His research focuses on new transition metal catalyzed cross-coupling reactions. Projects include exploring both the mechanism and synthetic capabilities of these reactions, with particular emphasis on the generation of phenylacetylenes for use in nanoelectronics and supramolecular chemistry. He is also interested in studying the pedagogy of organic chemistry.

Mio holds a B.S. in chemistry from the University of Detroit Mercy and a Ph.D. in organic chemistry from the University of Illinois at Urbana-Champaign. He was awarded a Mellon Fellowship to perform post-doctoral research and teaching at Macalester College (St. Paul, MN). Mio joined UDM's faculty in 2002.

Mark A. Benvenuto

Mark Benvenuto is a Professor of Chemistry at the University of Detroit Mercy, in the Department of Chemistry & Biochemistry. His research thrusts span a wide array of subjects, but include the use of energy dispersive X-ray fluorescence spectroscopy to determine trace elemental compositions of aquatic and land-based plant matter, food and dietary supplements, and medieval and ancient artifacts.

Benvenuto received a B.S. in chemistry from the Virginia Military Institute, and after several years in the Army, a Ph.D. in inorganic chemistry from the University of Virginia. After a post-doctoral fellowship at The Pennsylvania State University, he joined the faculty at the University of Detroit Mercy in 1993.

© 2014 American Chemical Society

Indexes

Author Index

Adams, M., 5
Akinbo, O., 149
Bachofer, S., 135
Benvenuto, M., 1, 67
Bilia, A., 23
Bouvier-Brown, N., 105
Donovan, W., 23
Farrell, J., 123
Field, K., 209
Goldman, A., 209

Latch, D., 193
Laviska, D., 209
Lawson, D., 87
Mio, M., 1, 67
Nickel, A., 123
Roberts-Kirchhoff, E., 1, 73
Smith, G., 23
Sparks, S., 209
Wheland, E., 23

Subject Index

A

Alameda Beltway site, 139
  eastern portion, 140*f*
  Marina Square Business Park and residential areas, 140*f*
  sampling design, schematic diagram, 141*f*
  XRF soil lead values, 144*t*
  XRF spectral data, 143*f*
  XRF spectral results, 142
Apathy
  toward civic engagement
    impact, 154
    manifestations, 152
    remedy, 154
    root causes, 153
  toward learning, contributing factors, and its global impact, 150

B

Bottled water analysis, 149
  addressing apathy toward learning and civic engagement, 155
    biplot of variables and factors after varimax rotation, 180*f*
    combining results of all water samples, 178
    comparison of bottled water, tap water, and well water samples, 170
    content and skill intended for each project, 164*t*
    contribution of factors to variability of samples, 176*f*
    correlations between variables and factors, 176*t*
    correlations between variables and factors after varimax rotation, 179*t*
    data quality objectives (DQO), 167*t*
    factor-variable correlations, 177*f*
    instrument accuracy using CRM-TMDW-B, 169*f*
    instrument calibration, 168*t*
    instrument calibration stability check, 169*f*
    instrument performance characteristics, 168
    instrument stability monitoring, 167*f*
    interrelationships, 157*f*
    learner as member of global community, 163*f*
    making sense of data using principal component analysis, 176
    municipal tap water from campus, 175
    municipal tap water from faculty member's home, 175
    panacea, 158
    preliminary implementation hybrid intervention, 165
    project-based learning, 158
    relative percent difference (RPD) of select samples, 170*t*
    results of student assessment, learning gains and teaching effectiveness, 181*f*
    sample Capstone project from analytical chemistry 1
    sample-factor correlations, 178*f*, 179*f*
    service-learning, 161
    well water from staff member living in suburb, 175
  apathy, 150
  concentrations of mineral elements
    bottled water samples, 171*t*
    municipal tap and well water samples, 173*t*
  recommendation for implementers
    avoiding mistakes of early implementers, 183
    handling failed experiment, 184
    project-based learning, considering major benefits, 185
  results and reflections on impacts of implementation of hybrid intervention
    impact on community, 182
    impact on instructor, 183
    impact on students, 182

E

Environmental chemistry, 102
Environmental connections and incorporation of service learning into chemistry courses, 2
Environmental justice in New Orleans and beyond
  additional classroom activities, 15
  Center for the Advancement of Teaching (CAT), 7

chemical plant accidents reported to LDEQ 2005-2013, 11t
community partner, Louisiana Bucket Brigade, 10
community project, 15
   informal blog post, 17
   neighborhoods, 16
   record, 16
course readings and other assignments, 12
   MSDS assignment, industrial chemicals, 13t
   specific writing prompts, 13
course sequence description, 6
development of course theme, 8
first-semester course, 7
open classroom discussion, 9
refinery accidents reported to LDEQ 2005-2013, 11t
second-semester course, 7
semester wrap-up, 18
student response and outcomes, 19
thought-provoking paper, 9
Environmental justice through atmospheric chemistry
   adding social context to environmental chemistry, 105
   direct local sampling, 109
   example exercises
      anthropogenic VOCs, 118f
      biogenic VOCs, 118f
      California Air Resources Board (ARB) ozone and NO_x concentrations in Los Angeles, 110
      direct gound VOC measurements, 117
      linear regression and correlation coefficients, 117t
      NASA data visualized in Google Earth, 116f
      NASA satellite data and Google Earth, 114
      ozone or NO_x and median household income, linear regression and correlation coefficients, 112t
      relationship between ambient ozone and NO_x concentrations, 114f
      trends of ambient ozone and NO_x concentrations, 113f
   hands-on learning, 107
   introduction to environmental justice and air pollution, 106
   publically available datasets, 108
   summary, 119

G

Green action through education (GATE)
   all significant ATSI pre/post findings for GATE and comparison LCs, 30t
   appendices
      assessment of water quality, chemist's approach, 36
      math and economics of water usage, 52
      mathematics of water usage, 48
   challenges of STEM reform, 24
   findings about student attitudes and habits of mind, 29
   student and staff ratings, 30t
   Indiana University-Purdue University Indianapolis (IUPUI), 24
   the model, 26
      LC block schedule (fall semester 2009 and 2010), 27t
      LC block schedule (fall semester 2012), 27t
      LC engagements and underlying STEM-specific learning objectives, 29t
      LC STEM-specific learning objectives and indicators, 28t
   overall project and goals
      build STEM faculty community, 26
      impact on students, 25
      implement educational innovation, 25
      offer STEM for all, 26
   SENCER (Science Education for New Civic Engagements and Responsibilities), 23
   student's portfolio, engagement and involvement with civic issue, 31
   students' comments, 31
   University of Akron (UA), 24
   water log activity, 32

I

Instrumental analysis at Seattle University, 193
   course description, 194
      academic learning outcomes, 195
      personal learning goals, 195
   discussion and summary, 204
   introduction, 194
   post-course survey question
      community partner's in-class presentation, 202f

mix of laboratory projects in class, 203f
student learning and motivation, 201f
students' connection to community, 203f
students' ideas for improving course, 204f
use of environmentally themed scientific literature activities, 202f
service-learning projects, introduction, 196
student feedback, summary
   numeric pre- and post-course surveys, 200
   qualitative student feedback, 200
   results of pre- and post-course surveys, 201f

L

Learning Enhanced through Experimental Design and Analysis with Rutgers (LEEDAR) program, 209 2014
   examples of student experiments, 221f
   experimental design, 219
   experimentation, 220
   overview, 218
   presentation of results and evaluations, 220
   discussion of automobile CO_2 emissions, 214f
   dry ice as learning tool, 213
   experimental design, 216
      general questions, 217f
   experimentation, 215
   fossil fuels, 214
      discussion of consumer products derived, 215f
   history and development, 211
   preliminary goals, 212
   preliminary laboratory experiment, 216f
   presentation slide showing CENTC institutions, 211f
   summary
      challenges and pitfalls, 223
      feedback, 224
      future directions, 226
      growth of program, 225
      service-learning through mentoring, 222
      statistics for LEEDAR Program from 2007-2013, 225f

M

Merit badge
   environmental connections and modern badge requirements, 68
   history, 68
   logistics of earning badge, 69
   planning for chemistry merit badge clinic
      BSA certified counselor, 70
      connecting to local scout troops through student members, 70
      designate spot or room for scout, 71
      dress rehearsal, 70
      enough safety goggles, 70
      extra mentor-volunteer, 70
      schedule event, 70

O

OMI. See Ozone monitoring instrument (OMI)
Ozone monitoring instrument (OMI), 114

S

Service learning, chemistry, and environmental connections
   incorporation of service learning into chemistry courses, 1
Service learning in environmental chemistry, 87
   amount of IR absorbing gas, 99
   assessment, 101
   climate change, 89
   climate models, 92
   CO_2 catches IR radiation, 92f
   CO_2 levels, 90
   common IR absorbing atmospheric gases, 98t
      planets Mars, 101t
      planets Venus, 100t
   developing model, group approach, 91
   earth model of intermediate complexity (EMIC), 93
   general circulation models (GCMs), 93
   global temperature model, 89
   modern climate models, 93
   range of students, 100
   service model environment, 89
   short-term weather predictions, 90
   Venus and Mars, 97

zero-dimensional energy balance model (0D-EBM), 93

T

Teach students importance of societal implications of nanotechnology
  incorporation of societal impacts of nanotechnology into course, 127
  Crichton portrayed nanotechnology, 129
  elements of societal impacts, 128
  interdisciplinary approach of nanotech course, 128
  reading and evaluating *Prey*, 129
  science fiction author, 129

Theme of National Chemistry Week
  activities for annual chemistry day, junior girl scouts, 76*f*
  annual chemistry day, 76
  Junior Girl Scouts National Leadership Journey series, 77
  theme, 77
  Chemistry 1025, 74
  civic engagement student pre-service survey, 82
  course and project description, 74
  "energy-now and forever", 73
  grading rubric for energy-themed posters, 80*t*
  Chemistry 1015 student scores, 78
  pre-service survey and post-service survey, 82*t*
  representative comments, 78
  scoring of student posters, 79*f*

U

Using service learning
  teach students importance of societal implications of nanotechnology
  National Nanotechnology Initiative (NNI), 124
  pedagogy supporting incorporation of service learning, 125
  service learning, specific goals and benefits, 124
  service learning assignment and course details, 131
  service learning to enhance societal impacts of nanotechnology, 125
  service learning to strengthen theme of societal impacts, 130

X

XRF soil screening at Alameda Beltway, 135
  highway 24 median soils analyzed for lead by XRF, 139*t*
  highway instructional site results, 139
  instrument calibration, 138

Z

Zero-dimensional energy balance model including IR active molecules
  atmospheric CO_2 levels measured at Mauna Loa, 96
  atmospheric IR adsorption model, 95
  carbon dioxide, 95
  chemical calorimetry, 96
  concentration of CO_2, 96
  earth's energy balance, 95
  energy balance equation (EBE), 94
  non-molecular approach, 94
  thermal energy, 94
  understanding of climate, 94
  vibrational mode of CO_2, 97